Chemistry
FOR
DUMMIES®
2ND EDITION

by John T. Moore, EdD
Professor of Chemistry,
Stephen F. Austin State University

WILEY

Wiley Publishing, Inc.

Chemistry For Dummies® 2nd Edition

Published by
Wiley Publishing, Inc.
111 River St.
Hoboken, NJ 07030-5774
www.wiley.com

Copyright © 2011 by Wiley Publishing, Inc., Indianapolis, Indiana

Published by Wiley Publishing, Inc., Indianapolis, Indiana

Published simultaneously in Canada

For general information on our other products and services, please contact our Customer Care Department within the U.S. at 877-762-2974, outside the U.S. at 317-572-3993, or fax 317-572-4002.

For technical support, please visit www.wiley.com/techsupport.

Wiley also publishes its books in a variety of electronic formats. Some content that appears in print may not be available in electronic books.

Library of Congress Control Number: 2011926326

ISBN: 978-1-118-00730-3

Manufactured in the United States of America

10 9 8 7 6 5 4 3 2

WILEY

About the Author

John T. Moore, EdD, grew up in the foothills of western North Carolina. He attended the University of North Carolina–Asheville where he received his bachelor's degree in chemistry. He earned his master's degree in chemistry from Furman University in Greenville, South Carolina. After a stint in the United States Army, he decided to try his hand at teaching. In 1971, he joined the chemistry faculty of Stephen F. Austin State University in Nacogdoches, Texas, where he still teaches chemistry. In 1985, he started back to school part-time, and in 1991 he received his doctorate in education from Texas A&M University.

John's area of specialty is chemical education. He has developed several courses for students planning on teaching chemistry at the high school level. In the early 1990s, he shifted his emphasis to training elementary education majors and in-service elementary teachers in hands-on chemical activities. He has received four Eisenhower grants for professional development of elementary teachers and has served as coeditor (along with one of his former students) of the "Chemistry for Kids" feature of *The Journal of Chemical Education*. He is the author of several books on chemistry and is coauthor of several more, including *Biochemistry For Dummies* and *Organic Chemistry II For Dummies*.

Although teaching has always been foremost in his heart, John found time to work part-time for almost five years in the medical laboratory of the local hospital and has been a consultant for a textbook publisher. He is active in a number of local, state, and national organizations, such as the Nacogdoches Kiwanis Club and the American Chemical Society.

John lives in the Piney Woods of East Texas with his wife Robin and their two dogs and two cats. He enjoys brewing his own beer and mead and making custom knife handles from exotic woods. He also loves to cook. His two boys, Jason and Matt, along with his daughter-in-law Sarah and two grandchildren, Zane and Sadie, remain in the mountains of North Carolina.

Dedication

This book is dedicated to those children, past, present, and future, who will grow to love chemistry, just as I have done. You may never make a living as a chemist, but I hope that you will remember the thrill of your experiments and will pass that enjoyment on to your children. This book is also dedicated to my family: my wife Robin, who encouraged me and put up with my foul moods close to deadlines; my two sons, Jason and Matthew; Jason's wife Sarah; and the two most wonderful grandchildren in the world, Zane and Sadie.

Author's Acknowledgments

I would not have had the opportunity to write this book without the encouragement of my agent, Grace Freedson. She took the time to answer my constant e-mails. I owe many thanks to the staff at Wiley, especially executive editor Lindsay Lefevere, senior project editor Chad Sievers, copy editor Caitie Copple, and technical editors Michael R. Mueller, PhD, and Todd Crabill, MA. Thanks also to my colleagues, who kept asking me how it was going, and especially Rich Langley, my friend and writing partner. And let me offer many thanks to all my students over the past 40 years, especially the ones who became teachers. I've learned from you, and I hope that you've learned from me.

Publisher's Acknowledgments

We're proud of this book; please send us your comments at http://dummies.custhelp.com. For other comments, please contact our Customer Care Department within the U.S. at 877-762-2974, outside the U.S. at 317-572-3993, or fax 317-572-4002.

Some of the people who helped bring this book to market include the following:

Acquisitions, Editorial, and Media Development

Senior Project Editor: Chad R. Sievers
(Previous Edition: Tim Gallan)

Acquisitions Editor: Lindsay Lefevere
(Previous Edition: Greg Tubach, Kathy Cox)

Copy Editor: Caitie Copple
(Previous Edition: Greg Pearson, Sandy Blackthorn)

Assistant Editor: David Lutton

Technical Editors: Michael R. Mueller, PhD, and Todd Crabill, MA

Editorial Manager: Michelle Hacker

Editorial Assistant: Rachelle Amick

Art Coordinator: Alicia B. South

Cover Photos: @istockphoto/jut13

Cartoons: Rich Tennant (www.the5thwave.com)

Composition Services

Project Coordinator: Sheree Montgomery

Layout and Graphics: Corrie Socolovitch, Christin Swinford

Proofreaders: Laura Albert, Laura Bowman

Indexer: BIM Indexing & Proofreading Services

Publishing and Editorial for Consumer Dummies

  Diane Graves Steele, Vice President and Publisher, Consumer Dummies

  Kristin Ferguson-Wagstaffe, Product Development Director, Consumer Dummies

  Ensley Eikenburg, Associate Publisher, Travel

  Kelly Regan, Editorial Director, Travel

Publishing for Technology Dummies

  Andy Cummings, Vice President and Publisher, Dummies Technology/General User

Composition Services

  Debbie Stailey, Director of Composition Services

Contents at a Glance

Table of Contents

Part IV: Environmental Chemistry: Benefits and Problems 265

Chapter 18: Cough! Cough! Hack! Hack! Air Pollution 267

Chapter 19: Examining the Ins and Outs of Water Pollution 279

Introduction

*Y*ou've passed the first hurdle in understanding a little about chemistry: You've picked up *Chemistry For Dummies,* 2nd Edition. I imagine that a large number of people looked at the title, saw the word *chemistry,* and bypassed it like it was covered in germs.

I don't know how many times I've been on vacation, struck up a conversation with someone, and been asked the dreaded question: "What do you do?"

"I'm a teacher," I reply.

"Really? And what do you teach?"

I steel myself, grit my teeth, and say in my most pleasant voice, "Chemistry."

I see The Expression, followed by, "Oh, I never took chemistry. It was too hard." Or "You must be smart to teach chemistry." Or "Goodbye!" If I were still in the dating scene, "Hi, I teach chemistry" would not be a good pick-up line!

I think a lot of people feel this way because they think that chemistry is too abstract, too mathematical, too removed from their real lives. But in one way or another, all of us do chemistry.

Remember making that baking soda and vinegar volcano as a child? That's chemistry. Do you cook or clean or use fingernail polish remover? All that is chemistry. I never had a chemistry set as a child, but I always loved science. My high school chemistry teacher was a great biology teacher but really didn't know much chemistry. But when I took my first chemistry course in college, the labs hooked me. I enjoyed seeing the colors of the solids coming out of solutions. I enjoyed *synthesis,* making new compounds. The idea of making something nobody else had ever made before fascinated me. I wanted to work for a chemical company, doing research, but then I discovered my second love: teaching.

Chemistry is sometimes called the central science (mostly by chemists), because in order to have a good understanding of biology or geology or even physics, you must have a good understanding of chemistry. Ours is a chemical world, and I hope that you enjoy discovering the chemical nature of it — and that afterward, you won't find the word *chemistry* so frightening.

About This Book

My goal with this book is not to make you into a chemistry major. My goal is simply to give you a basic understanding of some chemical topics that commonly appear in high school or college introductory chemistry courses. If you're taking a course, use this book as a reference in conjunction with your notes and textbook.

Simply watching people play tennis, no matter how intently you watch them, will not make you a tennis star. You need to practice. And the same is true with chemistry. It's *not* a spectator sport. If you're taking a chemistry course, then you need to practice and work on problems. I show you how to work certain types of problems — gas laws, for example — but use your textbook for practice problems. It's work, yes, but it really can be fun.

As I updated this second edition of *Chemistry For Dummies,* I reflected on what to include. I've enjoyed getting e-mails from people all over the world asking questions about the first edition or thanking me. However, looking at the overall feedback, I felt that I hadn't included quite enough about calculations and some other topics that students taking a college or high school–level class really needed. So in this second edition I beefed up the calculations and included some extra topics normally found in the first year of high school chemistry or the first semester of general chemistry in college. Overall, this edition will be more useful to those of you taking the chemistry course. For those of you who want some help with second-semester topics, hang in there and maybe, just maybe, you'll soon see *Chemistry II For Dummies* in your local bookstore.

Foolish Assumptions

I really don't know why you bought this book (or will buy it — in fact, if you're still in the bookstore and *haven't* bought it yet, buy two and give one as a gift), but I assume that you're taking (or retaking) a chemistry course or preparing to take a chemistry course. I also assume that you feel relatively comfortable with arithmetic and know enough algebra to solve for a single unknown in an equation. And I assume that you have a scientific calculator capable of doing exponents and logarithms.

And if you're buying this book just for the thrill of finding out about something different — with no plan of ever taking a chemistry course — I applaud you and hope that you enjoy this adventure. Feel free to skip those topics that don't hold your interest; for you, there will be no tests, only the thrill of increasing your knowledge about something new.

What Not to Read

I know you're a busy person and want to get just what you need from this book. Although I want you to read every single word I've written, I understand you may be on a time crunch. I keep the material to the bare bones, but I include a few sidebars. They're interesting reading (again, at least to me) but not really necessary for understanding the topic at hand, so feel free to skip them. This is *your* book; use it any way you want.

I mark some paragraphs with Technical Stuff icons. What I tell you in these paragraphs is more than you need to know, strictly speaking, but it may give you helpful or interesting detail about the topic at hand. If you want just the facts, you can skip these paragraphs.

How This Book Is Organized

I present this book's content in a logical progression of topics. But this doesn't mean you have to start at the beginning and read to the end of the book. Each chapter is self-contained, so feel free to skip around. Sometimes, though, you'll get a better understanding if you do a quick scan of a background section as you're reading. To help you find appropriate background sections, I've placed "see Chapter X for more information" cross-references here and there throughout the book.

Because I'm a firm believer in concrete examples, I also include lots of illustrations and figures with the text. They really help in the understanding of chemistry topics. And to help you with the math, I break up problems into steps so that you can easily follow exactly what I'm doing.

I've organized the topics in a logical progression — basically the same way I organize my courses for science and non-science majors. Following is an overview of each part of the book.

Part I: The Basic Concepts of Chemistry

In this part, I introduce you to the really basic concepts of chemistry. I define chemistry and show you where it fits among the other sciences (in the center, naturally). I show you the chemical world around you and explain why chemistry should be important to you. I also have a chapter (Chapter 2) devoted to chemical calculations. I show you how to use the factor label method of calculations, along with an introduction to the SI (metric) system. I also show you the three states of matter and talk about going from one state to another — and the energy changes that occur.

Besides covering the macroscopic world of things like melting ice, I cover the microscopic world of atoms. I explain the particles that make up the atom — protons, neutrons, and electrons — and show you where they're located in the atom.

I discuss how to use the periodic table, an indispensable tool for chemists. And I introduce you to the atomic nucleus, including the different subatomic particles. Finally, I introduce you to the wonderful world of gases. In fact, in the gas chapter, you can see so many gas laws (Boyle's law, Charles's law, Gay-Lussac's law, the combined gas law, the ideal gas law, Avogadro's law, and more) that you may feel like a lawyer when you're done. The material in these chapters gets you ready for additional topics in chemistry.

Part II: A Cornucopia of Chemical Concepts

In this part, you get into some really good stuff: chemical reactions. I give some examples of the different kinds of chemical reactions you may encounter and show you how to balance them. (You really didn't think I could resist that, did you?) I also introduce the mole concept. Odd name, yes, but the mole is central to your understanding of chemical calculations. It enables you to figure the amount of reactants needed in chemical reactions and the amount of product formed. I also talk about solutions and how to calculate their concentrations. And I explain why I leave the antifreeze in my radiator during the summer and why I add rock salt to the ice when I'm making ice cream.

This part gets into thermochemistry. Energy changes take place during chemical reactions. Some reactions give off energy (mostly in the form of heat), and some absorb energy in the form of heat. I show you how to figure how much heat is released. It may be enough to make you break out in a sweat. Finally, I tell you about acids and bases, things sour and things bitter. I discuss how to calculate their concentration and the pH of a solution.

Part III: Blessed Be the Bonds That Tie

I start off in this part talking about quantum theory, through which an electron can be represented by the properties of both particles and waves. In the first chapter, I throw certainties out the window and introduce you to probabilities. Then I explain bonding. I show you how table salt is made in Chapter 13, which covers ionic bonding, and I show you the covalent bonding of water in Chapter 14. I explain how to name some ionic compounds and how to draw Lewis structural formulas of some covalent ones. I even show you what some of the molecules look like. (Rest assured that I define all these techno-buzzwords on the spot, too.)

I also talk about periodic trends of the elements and intermolecular forces, those extremely important forces that give water its most unusual properties.

Part IV: Environmental Chemistry: Benefits and Problems

In this part, I discuss some environmental issues, specifically air and water pollution. I demonstrate what causes those pollutants and what chemistry can do to correct those problems. These issues, which are so often in the news, are among the most important problems society faces, and in order to evaluate possible solutions, you must have a little knowledge of chemistry. I hope that you don't get lost in the smog!

Finally, I introduce you to nuclear chemistry, with discussions about radioactivity, carbon-14 dating, fission, and fusion nuclear reactors.

Part V: The Part of Tens

In this part, I introduce you to ten great serendipitous chemical discoveries, ten great chemistry nerds (nerds rule!), and ten useful tips for passing Chem I. I started to put in my ten favorite chemistry songs, but I could only think of nine. Bummer. I also include a chapter on ten common chemicals used today to help you understand how basic chemistry affects daily life.

Icons Used in This Book

If you've read other *For Dummies* books, you'll recognize the icons used in this book, but here's the quickie lowdown for those of you who aren't familiar with them:

This icon gives you a tip on the quickest, easiest way to perform a task or conquer a concept. This icon highlights stuff that's good to know and stuff that'll save you time and/or frustration.

The Remember icon is a memory jog for those really important things you shouldn't forget.

I use this icon when safety in doing a particular activity, especially mixing chemicals, is described.

This icon points out different example problems you may encounter with the respective topic. I walk you through them step by step to help you gain confidence.

I don't use this icon very often because I keep the content pretty basic. But in those cases where I expand on a topic beyond the basics, I warn you with this icon. You can safely skip this material, but you may want to look at it if you're interested in a more in-depth description.

Where to Go from Here

Where to go from here is really up to you and your prior knowledge. If you're trying to clarify something specific, go right to that chapter and section. If you're a real novice, start with Chapter 1 and go from there. If you know a little chemistry, I suggest quickly reviewing Part I and then going on to Part II. Chapter 8 on the mole is essential, and so is Chapter 6 on gases.

If you're most interested in environmental chemistry, go on to Chapters 18 and 19. You really can't go wrong. I hope that you enjoy your chemistry trip.

Part I
The Basic Concepts of Chemistry

The 5th Wave   By Rich Tennant

"How we doin' over here? Anyone need their hydrogen atoms shaken up?"

In this part . . .

*I*f you are new to chemistry, it may seem a little frightening. I see students every day who've psyched themselves out by saying so often that they can't do chemistry. The good news: Anyone can figure out chemistry. Anyone can *do* chemistry. If you cook, clean, or simply exist, you're part of the chemical world.

I work with a lot of elementary school children, and they love science. I show them chemical reactions (vinegar plus baking soda, for example), and they go wild. And that's what I hope happens to you when you read this book and find out how interesting and important chemistry can be.

The chapters of Part I give you a background in chemistry basics. I show you how to do calculations and introduce you to the metric system. I tell you about matter and the states it can exist in, and I also talk a little about energy, including the different types and how it's measured. I discuss the microscopic world of the atom and its basic parts and explain how information about atoms is conveyed in the periodic table, the most useful tool for a chemist. And I cover the world of gases. This part takes you on a fun ride, so get your motor running!

Chapter 1

What Is Chemistry, and Why Do I Need to Know Some?

*1*f you're taking a course in chemistry, you may want to skip this chapter and go right to the area you're having trouble with. You already know what chemistry is — it's a course you have to pass. But if you bought this book to help you decide whether to take a course in chemistry or to have fun discovering something new, I encourage you to read this chapter. I set the stage for the rest of the book here by showing you what chemistry is, what chemists do, and why you should be interested in chemistry.

I really enjoy chemistry. It's far more than a simple collection of facts and a body of knowledge. I was a physics major when I entered college, but I was hooked when I took my first chemistry course. It seemed so interesting, so logical. I think it's fascinating to watch chemical changes take place, to figure out unknowns, to use instruments, to extend my senses, and to make predictions and figure out why they were right or wrong. The whole field of chemistry starts here — with the basics — so consider this chapter your jumping-off point. Welcome to the interesting world of chemistry.

Understanding What Chemistry Is

This whole branch of science is all about *matter,* which is anything that has mass and occupies space. *Chemistry* is the study of the composition and properties of matter and the changes it undergoes, including energy changes.

Science used to be divided into very clearly defined areas: If it was alive, it was biology. If it was a rock, it was geology. If it smelled, it was chemistry. If it didn't work, it was physics. In today's world, however, those clear divisions are no longer present. You can find biochemists, chemical physicists, geochemists, and so on. But chemistry still focuses on matter and energy and their changes.

A lot of chemistry comes into play with that last part — the changes matter undergoes. Matter is made up of either pure substances or mixtures of pure substances. The change from one substance into another is what chemists call a *chemical change,* or *chemical reaction,* and it's a big deal because when it occurs, a brand-new substance is created (see Chapter 3 for the nitty-gritty details).

So what are compounds and elements? Just more of the anatomy of matter. Matter is pure substances or mixtures of pure substances, and substances themselves are made up of either elements or compounds. (Chapter 3 dissects the anatomy of matter. And, as with all matters of dissection, it's best to be prepared — with a nose plug and an empty stomach.)

Distinguishing between Science and Technology

Science is far more than a collection of facts, figures, graphs, and tables. Science is a method for examining the physical universe. It's a way of asking and answering questions. However, in order for it to be called science, it must be testable. Being testable is what makes science different from faith.

For example, you may believe in UFOs, but can you test for their existence? How about matters of love? Does she love me? How much does she love me? Can I design a test to test and quantify that love? I think not. I have to accept that love on faith. It's not based in science, which is okay. Mankind has struggled with many great questions that science can't answer. Science is a tool that is useful in examining certain questions, but not all. You wouldn't use a front-end loader to eat a piece of pie, nor would you dig a ditch with a fork. Those are inappropriate tools for the task, just as science is an inappropriate tool for areas of faith.

Science is best described by the attitudes of scientists themselves: They're skeptical. They simply won't take another person's word for a phenomena — it must be testable. And they hold onto the results of their experiments tentatively, waiting for another scientist to disprove them. Scientists wonder, they question, they strive to find out *why,* and they experiment — they have exactly the same attitudes that most small children have before they grow up. Maybe this is a good definition of scientists — they are adults who've never lost that wonder of nature and the desire to know.

Technology, the use of knowledge toward a very specific goal, actually developed before science. Ancient peoples cooked food, smelted ores, made beer and wine by fermentation, and made drugs and dyes from plant material. Technology initially existed without much science. There were few theories and few true experiments. Reasoning was left to the philosophers. Eventually alchemy arose and gave chemistry its experimental basis. Alchemists searched for ways to turn other metals into gold and, in doing so, discovered many new chemical substances and processes, such as distillation. However, it wasn't until the 17th century that experimentation replaced serendipity (see the next section for a discussion of serendipity) and true science began.

Deciphering the Scientific Method

The *scientific method* is normally described as the way scientists go about examining the physical world around them. In fact, no one uses just one scientific method every time, but the one I cover here describes most of the critical steps scientists go through sooner or later. Figure 1-1 shows the different steps in the scientific method.

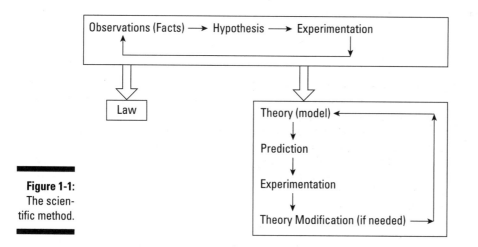

Figure 1-1:
The scientific method.

The following sections examine more in-depth what the scientific method is and how you can use it in all your studies, not just chemistry.

How the scientific method works

The way scientists are supposed to do their jobs is through the scientific method: a circular process that goes from observations to hypotheses to experiments and back to observations. These steps may lead in some cases to the creation of laws or theories.

To begin the scientific method, scientists make *observations* and note facts regarding something in the physical universe. The observations may raise a question or problem that the researcher wants to solve. He or she comes up with a *hypothesis,* a tentative explanation that's consistent with the observations (in other words, an educated guess). The researcher then designs an *experiment* to test the hypothesis. This experiment generates observations or facts that can then be used to generate another hypothesis or modify the current one. Then more experiments are designed, and the loop continues.

In good science, this loop of observations, hypothesis, and experimentation never ends. As scientists become more sophisticated in their scientific skills, think of better ways of examining nature, and build better and better instruments, their hypotheses are tested over and over. Conclusions that may appear to be scientifically sound today may be modified or even refuted tomorrow.

Besides continuing the loop, good experiments done with the scientific method may lead the researcher to propose a law or theory. A *law* is simply a generalization of what happens in the scientific system being studied. For example, the law of conservation of matter stated that matter is neither created nor destroyed. And like the laws that have been created for the judicial system, scientific laws sometimes have to be modified based on new facts. With the dawn of the nuclear age, scientists realized that in nuclear reactions a small amount of matter disappears and is converted to energy. So the law of conservation of matter was changed to read: In ordinary chemical reactions, matter is neither created nor destroyed.

A theory or model may also be proposed. A *theory* or *model* attempts to explain *why* something happens. It's similar to a hypothesis except that it has much more evidence to support it. What separates a theory from an opinion is that it has numerous experiments, many observations, and lots of data — in a nutshell, facts — supporting it.

The power of the theory or model is prediction. If the scientist can use the model to gain a good understanding of the system, then he or she can make predictions based on the model and then check them out with more experimentation. The observations from this experimentation can be used to refine or modify the theory or model, thus establishing another loop in the process. When does it end? Never. Again, as mankind develops more advanced instrumentation and ways of examining nature, scientists may find it necessary to modify our theories or models.

Science fairs and the scientific method

Suppose you're a high school student and your teacher is encouraging you to participate in the local science fair. You think and think about a project; you even buy *Science Fair Projects For Dummies* by Maxine Levaren (Wiley). A suggested experiment about energy content of nuts catches your eye and you decide to investigate which contains the most chemical energy — raw peanuts, roasted peanuts, or dry roasted peanuts. You think that nuts are roasted in oil so your hypothesis is that roasted peanuts contain more energy because of absorbed oil.

Now you have to design an experiment to test your hypothesis. You flip over to Chapter 10 on thermochemistry and read about calorimeters. You decide to make a calorimeter out of a couple of steel cans and a thermometer. You are careful to consider the variables involved — the mass of water, the mass of the nuts, and so on — and off you go to build your apparatus. You realize that you'll have to make several determinations on each type of peanut. You carefully and methodically collect your data, even doing an error analysis on the data.

After analyzing your data you may or may not have to modify your initial hypothesis. But then you begin to wonder if a cashew contains more energy per gram than a peanut — and what about all those other nuts in the grocery store? Your simple science fair project has generated more questions. And that is the road of the true scientists. Each investigation may answer some questions, but most probably will generate a lot more. Who knows, in 15 years you may find yourself working as a food chemist.

Many scientific discoveries are made through the scientific method. However, many discoveries are made by another process, called serendipity. *Serendipity* is an accidental discovery. The discoveries of penicillin, sticky notes, Velcro, radioactivity, Viagra, and so on were made by accident. But recognizing an accidental discovery takes a well-trained, disciplined, scientific mind. See Chapter 21 for a list of what I consider to be ten important serendipitous discoveries in chemistry.

How you can use the scientific method

Most people use the scientific method in their everyday lives without even thinking about it. You just think of it as tackling a problem logically. For example, suppose you buy that new HD TV and home theater system you've been wanting. You even buy a new CD changer so that you can listen to hours of music while studying. After unpacking and hooking everything up, you notice that you have no sound coming out of the left speakers when a CD is playing. You've identified a problem to investigate. Now you need to apply the scientific method to solve the problem. Here are some general steps to use:

1. **Develop a hypothesis about what you're studying.**

This hypothesis is an educated guess you make about what you think the end results will be. A hypothesis gives you an idea of what to expect, although after you conduct your experiments, you may determine the hypothesis is invalid.

For example, in the case of the dead left speakers, you may think that the problem lies with the CD changer, the receiver, or the cables connecting the two because everything else is working correctly. You form the hypothesis that something is wrong with the CD cables, that perhaps the left wire is broken or its connection is bad. You decide to experiment.

Identifying chemistry in the home

Chemistry is an important fact of everyday life. You can walk around your home to see all the chemistry-related things important to you. Check out chemistry in these rooms in your home:

- **Laundry room:** See that bottle of laundry detergent? Both the bottle and the detergent itself were made by chemists. You like those nice clean clothes, right? Without chemistry, you couldn't dress nearly as nice. Detergents contain a lot of things, including enzymes, brighteners, fillers, and so on, all of which chemists designed to make your clothes look good. Grab a bottle of bleach. Yep, made by chemists. Whether it be your clothes or your hair or wood pulp, chemists can get the color out of almost anything.

- **Closet:** If you wear clothes of something other than wool or cotton, you can thank a chemist and the chemical industry that discovered how to make those fibers.

- **Bathroom:** See that bar of soap? It was perfected by a chemist; otherwise, you would have to put up with grandma's harsh lye soap.

How about that toothpaste? There are a lot of ingredients in that simple product: colors, flavors, abrasives, thickeners, and fluoride, all designed by chemists. And I certainly hope that you use a deodorant. Every wonder what it contains? You can bet the formulation was developed by chemists.

What do you put on your skin? Probably lotions, powders, makeup, or cologne that was developed by chemists. And your hair — you wash it, curl it, straighten it, and color it, all with chemicals.

I know, it's enough to give you a headache. That aspirin you are getting ready to take is made by chemists, as well as the aceto-metaphin, ibuprofen, and so on. Chemicals are everywhere. Pull your hair out — and grow it back with a drug.

Chemists have given you the things you enjoy. Sometimes, problems arise in the process. Chemists have been and continue to be called upon to solve those problems.

2. **Conduct your experiment.**

Carefully design this experiment, with as many variables as possible being controlled. *Variables* are factors that can affect the outcome of the experiment. In chemistry, variables may be temperature, pressure, volume, and so on. (Controlling all the variables is very difficult when human beings are involved, which is why social-science experiments are so difficult.) In this example, the connections at both the CD player and the receiver are variables as well as the cable between the connections. You would only want to change one thing at a time. The simplest thing to do is to switch how the cable is connected at the CD unit. Just switch the right cable lead with the left one and vice versa. Suppose the left speakers are playing but the right set is dead. What does that tell you?

3. **Use the data and information from the experiment to generate a new hypothesis or modify the old one.**

Because the opposite speakers began malfunctioning when the CD cable connections were swapped, either the CD changer or the cable must be faulty, not the receiver. So you conduct another experiment, using a new set of cables. Thank goodness, everything is now playing just fine.

You may argue that the procedure you used was just common sense, but it really was the scientific method. In fact, I really do think of the scientific method as just good common sense.

Looking at the Branches of Chemistry

The general field of chemistry is so huge that it was originally subdivided into a number of different areas of specialization. But the different areas of chemistry now have a tremendous amount of overlap, just as there is among the various sciences. Here are the traditional fields of chemistry:

✔ **Analytical chemistry:** This branch is highly involved in the analysis of substances. Chemists from this field of chemistry may be trying to find out what substances are in a mixture (*qualitative analysis*) or how much of a particular substance is present (*quantitative analysis*) in something. Analytical chemists typically work in industry in product development or quality control. If a chemical manufacturing process goes wrong and is costing that industry hundreds of thousands of dollars an hour, that quality control chemist is under a lot of pressure to fix it and fix it fast. A lot of instrumentation is used in analytical chemistry. Chapters 7 through 9 cover a lot of the material that analytical chemists use.

✔ **Biochemistry:** This branch specializes in living organisms and systems. Biochemists study the chemical reactions that occur at the *molecular level* of an organism — the level where items are so small that people can't directly see them. Biochemists study processes such as digestion, metabolism, reproduction, respiration, and so on. Sometimes, distinguishing between a biochemist and a molecular biologist is difficult because they both study living systems at a microscopic level. However, a biochemist really concentrates more on the reactions that are occurring. For a good taste of biochemistry, see my book *Biochemistry For Dummies.*

✔ **Biotechnology:** This relatively new area of science is commonly placed with chemistry. It's the application of biochemistry and biology when creating or modifying genetic material or organisms for specific purposes. It's used in such areas as cloning and the creation of disease-resistant crops, and it has the potential for eliminating genetic diseases in the future. I also discuss this field in *Biochemistry For Dummies.*

✔ **Inorganic chemistry:** This branch is involved in the study of inorganic compounds such as salts. It includes the study of the structure and properties of these compounds. It also commonly involves the study of the individual elements of the compounds. Inorganic chemists would probably say that it is the study of everything except carbon, which they leave to the organic chemists.

✔ **Organic chemistry:** This is the study of carbon and its compounds. It's probably the most organized of the areas of chemistry — with good reason. There are millions of organic compounds, with thousands more discovered or created each year. Industries such as the polymer industry, the petrochemical industry, and the pharmaceutical industry depend on organic chemists.

✔ **Physical chemistry:** This branch figures out how and why a chemical system behaves as it does. Physical chemists study the physical properties and behavior of matter and try to develop models and theories that describe this behavior. Chapters 10 and 15 involve topics that physical chemists love.

Chemists, no matter what the type, all tend to examine the world around them in two ways — a macroscopic view and a microscopic view. The next sections take a look at these two viewpoints.

Macroscopic versus microscopic viewpoints

Most chemists that I know operate quite comfortably in two worlds. One is the *macroscopic* world that you and I see, feel, and touch. It's the world of stained lab coats — of weighing out things like sodium chloride to create things like hydrogen gas. The macroscopic realm is the world of experiments, or what some nonscientists call the "real world."

But chemists also operate quite comfortably in the *microscopic* world that you and I can't directly see, feel, or touch. Here, chemists work with theories and models. They may measure the volume and pressure of a gas in the macroscopic world, but they have to mentally translate the measurements into how close the gas particles are in the microscopic world.

Scientists often become so accustomed to slipping back and forth between these two worlds that they do so without even realizing it. An occurrence or observation in the macroscopic world generates an idea related to the microscopic world, and vice versa. You may find this flow of ideas disconcerting at first. But as you study chemistry, you'll soon adjust so that it becomes second nature.

Pure versus applied chemistry

In *pure chemistry,* chemists are free to carry out whatever research interests them — or whatever research they can get funded. They don't necessarily expect to find a practical application for their research at this point. The researchers simply want to know for the sake of knowledge. This type of research (often called *basic research*) is most commonly conducted at colleges and universities. Chemists use undergraduate and graduate students to help conduct the research. The work becomes part of the professional training of the student. The researchers publish their results in professional journals for other chemists to examine and attempt to refute. Funding is almost always a problem, because the experimentation, chemicals, and equipment are quite expensive.

In *applied chemistry,* chemists normally work for private corporations. Their research is directed toward a very specific short-term goal set by the company — product improvement or the development of a disease-resistant strain of corn, for example. Normally, more money is available for equipment and instrumentation with applied chemistry, but the chemists also have the pressure of meeting the company's goals.

These two types of chemistry, pure and applied, share the same basic differences as science and technology. In *science,* the goal is simply the basic acquisition of knowledge without any need for apparent practical application. Science is simply knowledge for knowledge's sake. *Technology* is the application of science toward a very specific goal.

Our society has a place for science *and* technology — likewise for the two types of chemistry. The pure chemist generates data and information that is then used by the applied chemist. Both types of chemists have their own sets of strengths, problems, and pressures. In fact, because of the dwindling federal research dollars, many universities are becoming much more involved in gaining patents, and they're being paid for technology transfers into the private sector.

Eyeing What You'll Do in Your Chemistry Class

I bet that somewhere along the way, you wondered what you would be doing in your chemistry class. Perhaps that was the motivation that led you to buy this book. The activities that you will do in class, especially the laboratory portion, are the very activities that professional chemists earn a living doing. You can group the activities of chemists (and chemistry students) into these major categories:

- ✔ **Chemists (and chemistry students) analyze substances.** They determine what is in a substance, how much of something is in a substance, or both. They analyze solids, liquids, and gases. They may try to find the active compound in a substance found in nature, or they may analyze water to see how much lead is present. (See Chapters 7 and 9.)

- ✔ **Chemists (and chemistry students) create, or *synthesize,* new substances.** They may try to make the synthetic version of a substance found in nature, or they may create an entirely new and unique compound. They may try to find a way to synthesize insulin. They may create a new plastic, pill, or paint. Or they may try to find a new, more efficient process to use for the production of an established product. (See Chapters 7 and 8.)

- ✔ **Chemists (and chemistry students) create models and test the predictive power of theories.** This area of chemistry is referred to as *theoretical chemistry.* Chemists who work in this branch of chemistry use computers to model chemical systems. Theirs is the world of mathematics and computers. Some of these chemists don't even own a lab coat. (See Chapters 6 and 15.)

- ✔ **Chemists (and chemistry students) measure the physical properties of substances.** They may take new compounds and measure the melting points and boiling points. They may measure the strength of a new polymer strand or determine the octane rating of a new gasoline. (See Chapter 10.)

What you can do with a chemistry degree

Although you're just into your first semester or year of chemistry, you may be envisioning a life in chemistry. You may be thinking that all chemists can be found deep in a musty lab, working for some large chemical company, but chemists hold a variety of jobs in a variety of places:

✔ **Quality control chemist:** These chemists analyze raw materials, intermediate products, and final products for purity to make sure that they fall within specifications. They may also offer technical support for the customer or analyze returned products. Many of these chemists often solve problems when they occur within the manufacturing process.

✔ **Industrial research chemist:** Chemists in this profession perform a large number of physical and chemical tests on materials. They may develop new products or work on improving existing products, possibly working with particular customers to formulate products that meet specific needs. They may also supply technical support to customers.

✔ **Sales representative:** Chemists may work as sales representatives for companies that sell chemicals or pharmaceuticals. They may call on their customers and let them know of new products being developed, or they may help their customers solve problems.

✔ **Forensic chemist:** These chemists analyze samples taken from crime scenes or analyze samples for the presence of drugs. They may also be called to testify in court as expert witnesses.

✔ **Environmental chemist:** These chemists may work for water purification plants, the Environmental Protection Agency, the Department of Energy, or similar agencies. This type of work appeals to people who like chemistry but also like to get out in nature. They often go out to sites to collect their own samples.

✔ **Preservationist of art and historical works:** Chemists work to restore paintings or statues, and sometimes they work to detect forgeries. With air and water pollution destroying works of art daily, these chemists preserve our heritage.

✔ **Chemical educator:** Chemists working as educators may teach physical science and chemistry in schools. University chemistry teachers often conduct research and work with graduate students. Chemists may even become chemical education specialists for organizations such as the American Chemical Society.

These professions are just a few that chemists may find themselves in. I didn't even get into law, medicine, technical writing, governmental relations, or consulting. Chemists are involved in almost every aspect of society. Some chemists even write books.

Chapter 2

Contemplating Chemical Calculations

Chemistry has a lot of calculations. But they're nothing you can't handle — they're arithmetic and simple algebra. To help you get a firm grasp of the calculations you encounter, you need to know a few important things.

You need to be familiar with the measurement system chemists use, the SI system (probably better known to you as the metric system). You also need to know a very useful way of setting up a problem — the unit conversion method. Along the way, you also need a good understanding of significant figures and rounding off. All in all, this chapter has a bunch of math, but hang in there. Mastering the basics here can help you as you venture through this book and through any chemistry courses you take.

Grasping the SI Measurement System

Much of the work chemists do involves measuring physical properties, such as the mass, volume, or length of a substance. Because chemists must be able to communicate their measurements to other chemists all over the world, they need to speak the same measurement language. This language is the SI system of measurement (from the French *Systeme International*), related to the metric system, which you hopefully have used before. Minor differences exist between the SI and metric systems, but for the most part, they're very similar.

This section lists the SI prefixes, base units for physical quantities in the SI system, and some useful SI-to-English measurement conversions.

Eyeing the basic SI prefixes

In order to be able to correctly use the SI system, you need to have a firm understanding of what each prefix means. The good news: The SI system is a decimal system. In other words, it's easy to use as long as you know the prefixes.

SI has base units for mass, length, volume, and so on, and prefixes modify the base units. For example, *kilo-* means 1,000; a kilogram is 1,000 grams, and a kilometer is 1,000 meters. Use Table 2-1 as a handy reference for the abbreviations and meanings of some selected various SI prefixes.

Table 2-1	SI (Metric) Prefixes	
Prefix	*Abbreviation*	*Meaning*
tera-	T	1,000,000,000,000 or 10^{12}
giga-	G	1,000,000,000 or 10^9
mega-	M	1,000,000 or 10^6
kilo-	k	1,000 or 10^3
hecto-	h	100 or 10^2
deka-	da	10 or 10^1
deci-	d	0.1 or 10^{-1}
centi-	c	0.01 or 10^{-2}
milli-	m	0.001 or 10^{-3}
micro-	μ	0.000001 or 10^{-6}
nano-	n	0.000000001 or 10^{-9}
pico-	p	0.000000000001 or 10^{-12}

Units of length

The base unit for length in the SI system is the *meter*. The exact definition of meter has changed over the years, but it's now defined as the distance that light travels in a vacuum in $\frac{1}{299,792,458}$ of a second. Here are some SI units of length:

1 millimeter (mm) = 1,000 micrometers (μm)

1 centimeter (cm) = 10 millimeters (mm)

1 meter (m) = 100 centimeters (cm)

1 kilometer (km) = 1,000 meters (m)

Some common English to SI system length conversions are

1 mile (mi) = 1.61 kilometers (km)

1 yard (yd) = 0.914 meters (m)

1 inch (in) = 2.54 centimeters (cm)

Units of mass

The base unit for mass in the SI system is the *kilogram*. It's the weight of the standard platinum-iridium bar found at the International Bureau of Weights and Measures. Here are some SI units of mass:

1 milligram (mg) = 1,000 micrograms (μg)

1 gram (g) = 1,000 milligrams (mg)

1 kilogram (kg) = 1,000 grams (g)

Some common English-to-SI-system mass conversions are

1 pound (lb) = 454 grams (g)

1 ounce (oz) = 28.4 grams (g)

1 pound (lb) = 0.454 kilograms (kg)

1 grain (gr) = 0.0648 grams (g)

1 carat (car) = 200 milligrams (mg)

Units of volume

In the SI system, volume is measured in base units called *cubic meters*. However, chemists normally use the *liter*, 0.001 m^3, to measure volume. Here are some SI units of volume:

1 milliliter (mL) = 1 cubic centimeter (cm^3) = 1,000 microliters (μL)

1 liter (L) = 1,000 milliliters (mL)

Some common English to SI system volume conversions are

1 quart (qt) = 0.946 liters (L)

1 pint (pt) = 0.473 liter (L)

1 fluid ounce (fl oz) = 29.6 milliliters (mL)

1 gallon (gal) = 3.78 liters (L)

Units of temperature

Kelvin is the base unit for temperature in the SI system. 0K is called *absolute zero,* the temperature at which all atomic/molecular motion ceases. Water freezes at 273K and boils at 373K. Following are the three major temperature conversion formulas:

Celsius to Fahrenheit: $°F = (9/5)°C + 32$

Fahrenheit to Celsius: $°C = (5/9)(°F − 32)$

Celsius to Kelvin: $K = °C + 273$

Units of pressure

The SI unit for pressure is the *pascal,* where 1 pascal equals 1 newton per square meter. (A *newton* is a unit of force equal to 1 kg·m/s^2.) But pressure can also be expressed in a number of different ways, so here are some common pressure conversions:

1 millimeter of mercury (mm Hg) = 1 torr

1 atmosphere (atm) = 760 millimeters of mercury (mm Hg) = 760 torr

1 atmosphere (atm) = 29.9 inches of mercury (in Hg)

1 atmosphere (atm) = 14.7 pounds per square inch (psi)

1 atmosphere (atm) = 101 kilopascals (kPa)

Units of energy

The SI unit for energy (heat being one form) is the *joule,* but most folks still use the metric unit of heat, the *calorie.* Here are some common energy conversions:

1 calorie (cal) = 4.184 joules (J)

1 nutritional (food) Calorie (Cal) = 1 kilocalorie (kcal) = 4,184 joules (J)

1 British thermal unit (BTU) = 252 calories (cal) = 1,053 joules (J)

Handling Really Big or Really Small Numbers

People who work in chemistry become quite comfortable working with very large and very small numbers. For example, when chemists talk about the number of sucrose molecules in a gram of table sugar, they're talking about a very, very large number. But when they talk about how much a single sucrose molecule weighs in grams, they're talking about a very, very small number. You can use regular longhand expressions in chemistry, but those measurements and equations become very bulky. To make working with really large and small numbers easier and quicker, chemists use exponential or scientific notation. The following sections take a closer look at how you can simplify work with large and small numbers in your chemistry studies.

Exploring exponential and scientific notation

In *exponential notation* a number is represented as a value raised to a power of ten. The decimal point can be located anywhere within the number as long as the power of ten is correct. In *scientific notation* the decimal point is always located between the first and second digit — and the first digit must be a number other than zero.

Suppose that you have an object that's 0.00125 meters in length. Express it in a variety of exponential forms.

$$0.00125 \text{ m} = \quad 0.0125 \times 10^{-1} \text{ m, or}$$

$$0.125 \times 10^{-2} \text{ m, or}$$

$$1.25 \times 10^{-3} \text{ m, or}$$

$$12.5 \times 10^{-4} \text{ m, and so on.}$$

All these forms are mathematically correct as numbers expressed in exponential notation. But in scientific notation the decimal point is placed so that only one digit other than zero is to the left of the decimal point. In the preceding example, the number expressed in scientific notation is 1.25×10^{-3} m. Most scientists express numbers in scientific notation.

Here are some powers of ten and the numbers they represent:

$1 \times 10^0 = 1$

$1 \times 10^1 = 1 \times 10 = 10$

$1 \times 10^2 = 1 \times 10 \times 10 = 100$

$1 \times 10^3 = 1 \times 10 \times 10 \times 10 = 1,000$

$1 \times 10^4 = 1 \times 10 \times 10 \times 10 \times 10 = 10,000$

$1 \times 10^5 = 1 \times 10 \times 10 \times 10 \times 10 \times 10 = 100,000$

$1 \times 10^{-1} = \frac{1}{10} = 0.1$

$1 \times 10^{-2} = \frac{1}{100} = 0.01$

$1 \times 10^{-3} = \frac{1}{1000} = 0.001$

$1 \times 10^{-10} = \frac{1}{10,000,000,000} = 0.0000000001$

Adding and subtracting

To add or subtract numbers in exponential or scientific notation, both numbers must have the same power of ten. If they don't, you must convert them to the same power. Here's an addition example:

$(1.5 \times 10^3 \text{ g}) + (2.3 \times 10^2 \text{ g}) =$

$(15 \times 10^2 \text{ g}) + (2.3 \times 10^2 \text{ g}) =$

17.3×10^2 g (exponential notation) =

1.73×10^3 g (scientific notation)

Subtraction is done exactly the same way.

Multiplying and dividing

To multiply numbers expressed in exponential notation, multiply the coefficients (the numbers) and add the exponents (powers of ten):

$(9.25 \times 10^{-2} \text{ m}) \times (1.37 \times 10^{-5} \text{ m}) =$

$(9.25 \times 1.37) \times 10^{(-2 + -5)} \text{ m}^2 =$

$12.7 \times 10^{-7} \text{ m}^2 =$

$1.27 \times 10^{-6} \text{ m}^2$

To divide numbers expressed in exponential notation, divide the coefficients and subtract the exponent of the denominator from the exponent of the numerator:

$$(8.27 \times 10^5 \, g) \div (3.25 \times 10^3 \, mL) =$$

$$(8.27 \div 3.25) \times 10^{5-3} \, g/mL =$$

$$2.54 \times 10^2 \, g/mL$$

Raising a number to a power

To raise a number in exponential notation to a certain power, raise the coefficient to the power and then multiply the exponent by the power:

$$(4.33 \times 10^{-5} \, cm)^3 = (4.33)^3 \times 10^{-5 \times 3} \, cm^3 = 81.2 \times 10^{-15} \, cm^3 = 8.12 \times 10^{-14} \, cm^3$$

Relying on a calculator

Scientific calculators take a lot of drudgery out of doing calculations. They enable you to spend more time thinking about the problem itself.

You can use a calculator to add and subtract numbers in exponential notation without first converting them to the same power of ten. The only thing you need to be careful about is entering the exponential number correctly.

I assume that your calculator has a key labeled *EXP*. EXP stands for $\times 10^x$. After you press the EXP key, you enter the power. For example, to enter the number 6.25×10^3, you type 6.25, press the EXP key, and then type 3. As for the negative exponent, if you want to enter the number 6.05×10^{-12}, you type 6.05, press the EXP key, type 12, and then press the +/– key.

When using a scientific calculator, *don't* enter the $\times 10$ part of your exponential number. Press the EXP key to enter this part of the number.

Deciphering the Difference between Accuracy and Precision

Whenever you make measurements, you must consider two factors. *Accuracy* is how well the measurement agrees with the accepted or true value. *Precision* is how well a set of measurements agree with each other. In chemistry, measurements should be *reproducible;* that is, they must have a high degree of precision. Most of the time chemists make several measurements and average them. The closer these measurements are to each other, the more confidence chemists have in their measurements. However, you also want the measurements to be accurate, very close to the correct answer. However, many times you don't know beforehand anything about the correct answer; therefore, you have to rely on precision as your guide.

For example, suppose you ask four lab students to make three measurements of the length of the same object. Their data follows:

	Student 1	Student 2	Student 3	Student 4
Trial 1	27.77 cm	27.30 cm	27.55 cm	27.30 cm
Trial 2	27.30 cm	27.60 cm	27.55 cm	27.29 cm
Trial 3	27.56 cm	27.97 cm	27.53 cm	27.31 cm
Average	27.54 cm	27.62 cm	27.54 cm	27.30 cm

The accepted length of the object is 27.55 cm. Which of these students deserves the higher lab grade? Both students 1 and 3 have values close to the accepted value, if you just consider their average values. (The average, found by summing the individual measurements and dividing by the number of measurements, is normally considered to be more useful than any individual value.) Both students 1 and 3 have made *accurate* determinations of the length of the object. The average values determined by students 2 and 4 are not very close to the accepted value, so that their values are not considered to be accurate.

However, if you examine the individual determinations for students 1 and 3, you notice a great deal of variation in the measurements of student 1. The measurements don't agree with each other very well; their precision is low even though the accuracy is good. The measurements by student 3 agree well with each other; both precision and accuracy are good. Student 3 deserves a higher grade than student 1.

Neither student 2 nor student 4 has average values close to the accepted value; neither determination is very accurate. However, student 4 has values that agree closely with each other; the precision is good. This student probably had a consistent error in their measuring technique. Student 2 had neither good accuracy nor precision. The accuracy and precision of the four students is summarized below.

	Accuracy	Precision
Student 1	High	Low
Student 2	Low	Low
Student 3	High	High
Student 4	Low	High

Usually, measurements with a high degree of precision are also somewhat accurate. Because the scientists or students don't know the accepted value beforehand, they strive for high precision and hope that the accuracy will also be high. This was not the case for student 4.

Using the Unit Conversion Method to Solve Problems

You may find that sometimes you're unclear on how to actually set up chemistry problems in order to solve them. A scientific calculator handles the math, but it can't tell you what you need to multiply or what you need to divide.

That's why you need to know about the *unit conversion method,* which is sometimes called the *factor label method.* It can help you set up chemistry problems and calculate them correctly. Two basic rules are associated with the unit conversion method:

- ✔ **Rule 1:** Always write the unit and the number associated with the unit. Rarely in chemistry will you have a number without a unit. Pi is the major exception that comes to mind.

- ✔ **Rule 2:** Carry out mathematical operations *with* the units, canceling them until you end up with the unit you want in the final answer. In every step, you must have a correct mathematical statement.

Suppose that you have an object traveling at 75 miles per hour, and you want to calculate its speed in kilometers per second.

To solve this calculation using the unit conversion method, follow these steps:

1. **Write down what you start with:**

   $$\frac{75 \text{ mi}}{\text{hr}}$$

   Note that per Rule 1, the equation shows the unit and the number associated with it.

2. **Convert miles to feet, canceling the unit of miles per Rule 2:**

   $$\frac{75 \ \cancel{\text{mi}}}{\text{hr}} \times \frac{5,280 \text{ ft}}{1 \ \cancel{\text{mi}}}$$

3. **Convert feet to inches:**

   $$\frac{75 \ \cancel{\text{mi}}}{\text{hr}} \times \frac{5,280 \ \cancel{\text{ft}}}{\cancel{\text{mi}}} \times \frac{12 \text{ in}}{1 \ \cancel{\text{ft}}}$$

4. **Convert inches to centimeters:**

$$\frac{75 \text{ mi}}{\text{hr}} \times \frac{5{,}280 \text{ ft}}{\text{mi}} \times \frac{12 \text{ in}}{1 \text{ ft}} \times \frac{2.54 \text{ cm}}{1 \text{ in}}$$

5. **Convert centimeters to meters:**

$$\frac{75 \text{ mi}}{\text{hr}} \times \frac{5{,}280 \text{ ft}}{\text{mi}} \times \frac{12 \text{ in}}{1 \text{ ft}} \times \frac{2.54 \text{ cm}}{1 \text{ in}} \times \frac{1 \text{ m}}{100 \text{ cm}}$$

6. **Convert meters to kilometers:**

$$\frac{75 \text{ mi}}{\text{hr}} \times \frac{5{,}280 \text{ ft}}{\text{mi}} \times \frac{12 \text{ in}}{1 \text{ ft}} \times \frac{2.54 \text{ cm}}{1 \text{ in}} \times \frac{1 \text{ m}}{100 \text{ cm}} \times \frac{1 \text{ km}}{1{,}000 \text{ m}}$$

7. **Stop and stretch.**

8. **Now convert hours to minutes in the denominator of the original fraction:**

$$\frac{75 \text{ mi}}{\text{hr}} \times \frac{5{,}280 \text{ ft}}{\text{mi}} \times \frac{12 \text{ in}}{1 \text{ ft}} \times \frac{2.54 \text{ cm}}{1 \text{ in}} \times \frac{1 \text{ m}}{100 \text{ cm}} \times \frac{1 \text{ km}}{1{,}000 \text{ m}} \times \frac{1 \text{ hr}}{60 \text{ min}}$$

9. **Convert minutes to seconds:**

$$\frac{75 \text{ mi}}{\text{hr}} \times \frac{5{,}280 \text{ ft}}{\text{mi}} \times \frac{12 \text{ in}}{1 \text{ ft}} \times \frac{2.54 \text{ cm}}{1 \text{ in}} \times \frac{1 \text{ m}}{100 \text{ cm}} \times \frac{1 \text{ km}}{1{,}000 \text{ m}} \times \frac{1 \text{ hr}}{60 \text{ min}} \times \frac{1 \text{ min}}{60 \text{ s}}$$

10. **Do the math to get the answer now that you have the units of kilometers per second (km/s):**

   0.033528 km/s

11. **Round off your answer to the correct number of significant figures.**

   The next section gives you details on how to do so. The rounded-off answer to this problem is

   0.034 km/s or 3.4×10^{-2} km/s

Note that although the setup of the preceding example is correct, it's certainly not the only correct setup. Depending on what conversion factors you know and use, there may be many correct ways to set up a problem and get the correct answer. Check out another example to ensure you have this method down pat.

Suppose that you have an object with an area of 35 inches squared, and you want to figure out the area in meters squared.

Follow these easy steps to make this calculation.

1. **Write down what you start with:**

$$\frac{35 \text{ in}^2}{1}$$

2. Convert from inches to centimeters.

You have to cancel inches *squared.* You must square the inches in the new fraction, and if you square the unit, you have to square the number also. And if you square the denominator, you have to square the numerator, too:

$$\frac{35 \ \cancel{in}^2}{1} \times \frac{(2.54 \ cm)^2}{(1 \ \cancel{in})^2}$$

3. Convert from centimeters squared to meters squared in the same way:

$$\frac{35 \ \cancel{in}^2}{1} \times \frac{(2.54 \ \cancel{cm})^2}{(1 \ \cancel{in})^2} \times \frac{(1 \ m)^2}{(100 \ \cancel{cm})^2}$$

4. Now that you have the units of meters squared (m^2), do the math to get your answer:

0.0225806 m²

5. Round off your answer to the correct number of significant figures (see the next section for details), getting:

$0.023 \ m^2$ or $2.3 \times 10^{-2} \ m^2$

One of the concepts students often see in the introductory material of most chemistry texts is density. *Density* is defined as the mass per volume (density = mass/volume). I now show you how to use this definition in the unit conversion method.

Find the density, in g/cm³, of a sample of a liquid, given that 2.00 ft³ of the liquid weighs 97.5 pounds.

To solve this problem, stick to these steps:

1. Pull the information out of this problem:

Volume = 2.00 ft³   Mass = 97.5 lbs   Density = ? g/cm³

2. Put the mass and volume measurement into the density formula.

$$\frac{97.5 \ lb}{2.00 \ ft^3}$$

3. Convert from pounds to grams:

$$\frac{97.5 \ \cancel{lb}}{2.00 \ ft^3} \times \frac{453.6 \ g}{1 \ \cancel{lb}}$$

4. Convert from cubic feet to cubic inches:

$$\frac{97.5 \ \cancel{lb}}{2.00 \ \cancel{ft}^3} \times \frac{453.6 \ g}{1 \ \cancel{lb}} \times \frac{(1 \ \cancel{ft})^3}{(12 \ in)^3}$$

If you raise a unit to a power, you must also raise the number to that same power.

5. Convert from cubic inches to cubic centimeters:

$$\frac{97.5 \ \cancel{lb}}{2.00 \ \cancel{ft}^3} \times \frac{453.6 \ g}{1 \ \cancel{lb}} \times \frac{(1 \ \cancel{ft})^3}{(12 \ \cancel{in})^3} \times \frac{(1 \ \cancel{in})^3}{(2.54 \ cm)^3}$$

6. Do the arithmetic, getting:

Density = $0.780896 \ g/cm^3$ = $0.781 \ g/cm^3$

With a little practice, you'll really like and appreciate the unit conversion method. It even got me through my introductory physics course.

Knowing How to Handle Significant Figures

Significant figures (no, I'm not talking about supermodels) are the number of digits that you report in the final answer of the mathematical problem you're calculating. If I told you that one student determined the density of an object to be 2.3 g/mL and another student figured the density of the same object to be 2.272589 g/mL, I bet that you'd believe that the second figure was the result of a more accurate experiment. You may be right, but then again, you may be wrong. You have no way of knowing whether the second student's experiment was more accurate unless both students obeyed the significant figure convention.

The number of digits that you report in your final answer gives the reader some information about how accurately you made the measurements. The number of the significant figures is limited by the accuracy of the measurement.

This section shows you how to determine the number of significant figures in a number, how to determine how many significant figures you need to report in your final answer, and how to round off your answer to the correct number of significant figures.

Comparing numbers: Exact and counted versus measured

If I ask you to count the number of automobiles that you and your family own, you can do it without any guesswork involved. Your answer may be 0,

1, 2, or 10, but you know exactly how many autos you have. Those numbers are what are called *counted numbers*. If I ask you how many inches are in a foot, your answer will be 12. That number is an *exact number* — it's exact by definition. Another exact number is the number of centimeters per inch, 2.54. In both exact and counted numbers, you have no doubt what the answer is. When you work with these types of numbers, you don't have to worry about significant figures.

Now suppose that I ask you and four of your friends to individually measure the length of an object as accurately as you possibly can with a meter stick. You then report the results of your measurements: 2.67 meters, 2.65 meters, 2.68 meters, 2.61 meters, and 2.63 meters. Which of you is right? You are all within experimental error. These measurements are *measured numbers*, and measured values always have some error associated with them. You determine the number of significant figures in your answer by your least reliable measured number, and I tell you how in the following section.

Determining the number of significant figures in a measured number

To determine the number of significant figures, or *sig. figs.*, in a measured number, follow these rules:

- **Rule 1:** All nonzero digits are significant. All numbers, one through nine, are significant, so 676 contains three sig. figs., 5.3×10^5 contains two, and 0.2456 contains four. The zeroes are the only numbers you have to worry about.

- **Rule 2:** All the zeroes between nonzero digits are significant. For example, 303 contains 3 sig. figs., 425,003,704 contains nine, and 2.037×10^{-6} contains four.

- **Rule 3:** All zeros to the left of the first nonzero digit are *not* significant. For example, 0.0023 contains two sig. figs. and 0.0000050023 (which would be expressed in scientific notation as 5.0023×10^{-6}) contains five.

- **Rule 4:** Zeroes to the right of the last nonzero digit are significant if a decimal point is present. For example, 3,030.0 contains five sig. figs., 0.000230340 contains six, and 6.30300×10^7 also contains six sig. figs.

- **Rule 5:** Zeroes to the right of the last nonzero digit are *not* significant if no decimal point is present. (Actually, a more correct statement is that you really don't know about those zeroes if a decimal point isn't present. You would have to know something about how the value was measured. But most scientists use this rule as a convention.) For example, 72,000 contains two sig. figs. and 50,500 contains three.

Reporting the correct number of significant figures

In general, the number of significant figures that you report in your calculation is determined by the *least-precise* measured value. What values qualify as the least-precise measurement varies depending on the mathematical operations involved.

Addition and subtraction

In addition and subtraction, report your answer to the number of decimal places used in the number that has the fewest decimal places. For example, suppose you're adding the following amounts:

2.675 g + 3.25 g + 8.872 g + 4.5675 g

Your calculator will show 19.3645, but you round off to the hundredths place based on the 3.25, which has the fewest number of decimal places. You round the figure off to 19.36. (See the following section to find out the rules for rounding.)

Multiplication and division

In multiplication and division, you report the answer to the same number of significant figures as the number that has the *fewest* significant figures. Remember that counted and exact numbers don't count in the consideration of significant numbers. For example, suppose that you are calculating the density in grams per liter of an object that weighs 25.3573 (6 sig. figs.) grams and has a volume of 10.50 milliliters (4 sig. figs.). The setup looks like this:

$$\frac{25.3573 \text{ g}}{10.5 \text{ mL}} \times \frac{1,000 \text{ mL}}{1 \text{ L}}$$

Your calculator will read 2,414.981000. You have six significant figures in the first number and four in the second number (the 1,000 mL/L does not count because it is an exact conversion). You should have four significant figures in your final answer, so round the answer off to 2,415 g/L.

Rounding off numbers

When rounding off numbers, use the following rules:

- **Rule 1:** If the first number to be dropped is 5 or greater, drop it and all the numbers that follow it, and increase the last retained number by 1.

  For example, suppose that you want to round off 237.768 to four significant figures. You drop the 6 and the 8. The 6, the first dropped number, is greater than 5, so you increase the retained 7 to 8. Your final answer is 237.8.

- **Rule 2:** If the first number to be dropped is less than 5, drop it and all the numbers that follow it, and leave the last retained number unchanged.

  If you're rounding 2.35427 to three significant figures, you drop the 4, the 2, and the 7. The first number to be dropped is 4, which is less than 5. The 5, the last retained number, stays the same. So you report your answer as 2.35.

Only round off your final answer. Do not round off any values used in a calculation.

Chapter 3

Matter and Energy

. . . .

In This Chapter

▶ Understanding the states of matter and their changes

▶ Differentiating between pure substances and mixtures

▶ Examining the properties of chemical substances

▶ Discovering the different types of energy

▶ Measuring the energy in chemical bonds

. . . .

Walk into a room and turn on the light. Look around — what do you see? You may see a table, some chairs, a lamp, a computer humming away. But really all you see is matter and energy. There are many kinds of matter and many kinds of energy, but when all is said and done, you're left with these two things — matter and energy. Scientists used to believe that these two were separate and distinct, but now they realize that they're linked.

In this chapter, I cover the two basic components of the universe, matter and energy. You examine the different states of matter and what happens when matter goes from one state to another. I show you how the metric system is used to make matter and energy measurements, and I also discuss the different types of energy and the way energy is measured.

Looking at the Facts of Matter

Look around you. All the stuff you see — your chair, the water you're drinking, the paper this book is printed on — is matter. Matter is anything that has mass and occupies space. It's the material part of the universe. (Later in this chapter I introduce you to energy, the other part of the universe.)

Whenever scientists contemplate matter, they do so in one of two ways — with a macroscopic viewpoint or with a microscopic one. The following gives you a basic overview of each definition:

✔ When scientists view matter in a *macroscopic* way, they're considering matter as the "stuff" they can physically observe — a lump of coal, 5 pounds of sugar, a pinch of salt. The macroscopic world is the world you and I can directly observe through our senses. Most people use this viewpoint when looking at the world around them.

✔ The chemist can switch perspective to a microscopic one. The *microscopic* level isn't just what you can observe through a microscope. It goes far beyond that. It's the level of individual particles, such as carbon atoms, sugar molecules, and sodium and chloride ions. (Don't worry, I talk about all those things in upcoming chapters.) The microscopic view is the world of scientists' theories and models.

Scientists can switch back and forth between these viewpoints without even thinking, and my hope is that you can grow to feel more comfortable with viewing matter in both these ways. Whichever perspective you use, matter can exist in one of three states: solid, liquid, and gas.

Solids

At the macroscopic level, the level at which you directly observe with your senses, a solid has a definite shape and occupies a definite volume. Think of an ice cube in a glass — it's a solid. You can easily weigh the ice cube and measure its volume. At the microscopic level (where items are so small that people can't directly observe them), the particles that make up the ice are very close together and aren't moving around very much (see Figure 3-1a).

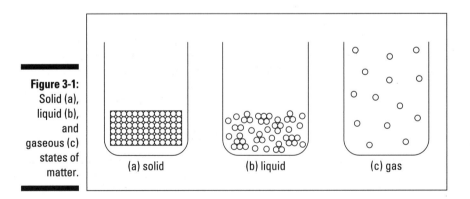

Figure 3-1:
Solid (a),
liquid (b),
and
gaseous (c)
states of
matter.

(a) solid   (b) liquid   (c) gas

The particles that make up the ice (also known as *water molecules*) are close together and have little movement because, as in many solids, the particles are pulled into a rigid, organized structure of repeating patterns called a *crystal lattice*. The particles that are contained in the crystal lattice are still moving, but barely — it's more of a slight vibration. Depending on the particles, this crystal lattice may be of different shapes.

Liquids

When an ice cube melts, it becomes a liquid. Unlike solids, liquids have no definite shape, but they do have a definite volume, just like solids do. For example, a cup of water in a tall, skinny glass has a different shape than a cup of water in a pie pan, but in both cases, the volume of water is the same — one cup. Why? The particles in liquids are not really much farther apart than the particles in solids, but they're moving around much more (refer to Figure 3-1b).

Some of the particles in liquids may be near each other, clumped together in small groups. Because the particles are moving much faster in liquids, the attractive forces among them aren't as strong as they are in solids — which is why liquids don't have a definite shape. However, these attractive forces are strong enough to keep the substance confined in one large mass — a liquid — instead of going all over the place.

Gases

If you heat water, you can convert it to *steam,* the gaseous form of water. A gas has no definite shape and no definite volume. In a gas, particles are much farther apart than they are in solids or liquids (refer to Figure 3-1c), and they're moving relatively independent of each other. Because of the distance between the particles and the independent motion of each of them, the gas expands to fill the area that contains it (and thus it has no definite shape).

Because a great deal of distance separates gas particles, you can easily compress a gas but not a solid — or, to a certain extent, a liquid where the particles are very close together. If you hold a balloon and squeeze, you can actually force those gas particles closer together because of all the empty space between the gas particles. Check out Chapter 6 for more information on gases.

Ice in Alaska, Water in Texas: Changing States of Matter

When a substance goes from one state of matter to another, the process is a *change of state*. Some rather interesting things, which I describe in the following sections, occur during this process.

I'm melting away! Oh, what a world!

Imagine taking a big chunk of ice out of your freezer and putting it into a large pot on your stove. If you measure the temperature of that chunk of ice, you may find it to be –5 degrees Celsius or so. If you take temperature readings while heating the ice, you find that the temperature of the ice begins to rise as the heat from the stove causes the ice particles to begin vibrating faster and faster in their crystal lattice. After a while, some of the particles move so fast that they break free of the lattice, and the crystal lattice (which keeps a solid solid) eventually breaks apart. The solid begins to go from a solid state to a liquid state — a process called *melting*. The temperature at which melting occurs is called the *melting point (mp)* of the substance. The melting point for ice is 32 degrees Fahrenheit, or 0 degrees Celsius.

If you watch the temperature of ice as it melts, you see that the temperature remains steady at 0 degrees Celsius until all the ice has melted. During this change of state *(phase change),* the temperature remains constant; the energy that you're adding goes to work breaking the lattice, freeing the particles as a liquid. The liquid contains more energy than the ice (because the particles in liquids move faster than the particles in solids), even if both the ice and liquid water are at the same temperature. This is true at all phase changes — during the phase change itself the temperature remains constant.

Boiling point

If you heat a pot of cool water (or if you continue to heat the pot of now-melted ice mentioned in the preceding section), the temperature of the water rises and the particles move faster and faster as they absorb the heat. The temperature rises until the water reaches the next change of state — boiling. As the particles move faster and faster as they heat up, they begin to break the attractive forces between each other and move freely as steam — a gas. The process by which a substance moves from the liquid state to the gaseous state is called *boiling*. The temperature at which a liquid begins to boil is called the *boiling point (bp)*. The bp is dependent on atmospheric pressure, but for water at sea level, it's 212 degrees Fahrenheit, or 100 degrees Celsius. The temperature of the boiling water remains constant until all the water has been converted to steam.

You can have both water and steam at 100 degrees Celsius. They have the same temperature, but the steam has a lot more energy (because the particles move independently and pretty quickly). Because steam has more energy, steam burns are normally a lot more serious than boiling water burns — loads more energy is transferred to your skin. I was reminded of this one morning while trying to iron a wrinkle out of a shirt that I was still wearing. My skin and I can attest — steam contains a *lot* of energy!

I can summarize the process of water changing from a solid to a liquid in this way:

ice → water → steam

Because the basic particle in ice, water, and steam is the water molecule (written as H_2O), the same process can also be shown as

$H_2O(s) \rightarrow H_2O(l) \rightarrow H_2O(g)$

Here the *(s)* stands for solid, the *(l)* stands for liquid, and the *(g)* stands for gas. This second depiction is much better, because unlike H_2O, most chemical substances don't have different names for the solid, liquid, and gas forms.

Freezing point

If you cool a gaseous substance, you can watch the phase changes that occur. The phase changes are

✔ **Condensation:** Going from a gas to a liquid

✔ **Freezing:** Going from a liquid to a solid

The gas particles have a high amount of energy, but as they're cooled, that energy is reduced. The attractive forces then have a chance to draw the particles closer together, forming a liquid. This process is called *condensation*. The particles are now in clumps (as is characteristic of particles in a liquid state), but as more energy is removed by cooling, the particles start to align themselves, and a solid is formed. This is known as *freezing*. The temperature at which this occurs is called the *freezing point (fp)* of the substance.

The freezing point is the same as the melting point — it's the temperature at which the liquid is able to become a solid or a solid becomes a liquid. In the same fashion, the condensation temperature is the same as the boiling point.

You can represent water changing states from a gas to a solid like this:

$$H_2O(g) \rightarrow H_2O(l) \rightarrow H_2O(s)$$

Sublimate this!

Most substances go through the logical progression from solid to liquid to gas as they're heated — or vice versa as they're cooled. But a few substances go directly from the solid to the gaseous state without ever becoming a liquid. Scientists call this process *sublimation*. Dry ice — solid carbon dioxide, written as $CO_2(s)$ — is the classic example of sublimation. (Mothballs and certain solid air fresheners also go through the process of sublimation.) You can see dry ice particles becoming smaller as the solid begins to turn into a gas, but no liquid is formed during this phase change. (If you've seen dry ice, then you remember that a white cloud usually accompanies it — magicians and theater productions often use dry ice for a cloudy or foggy effect. The white cloud you normally see isn't the carbon dioxide gas — the gas itself is colorless. The white cloud is the condensation of the water vapor in the air due to the cold of the dry ice.)

The process of sublimation is represented as

$$CO_2(s) \rightarrow CO_2(g)$$

The reverse of sublimation is *deposition* — going directly from a gaseous state to a solid state. One method of purifying solid iodine is to heat it gently. It sublimates and then it will undergo deposition on a cooler surface. Many times that cooler surface is a piece of laboratory apparatus that resembles a large test tube full of ice. It's commonly called a *cold finger*.

Classifying Pure Substances and Mixtures

One of the basic processes in science is classification. As I discuss in the preceding section, chemists can classify matter as solid, liquid, or gas. But matter can be classified in other ways as well. In this section, I discuss how all matter can be classified as either a pure substance or a mixture (see Figure 3-2).

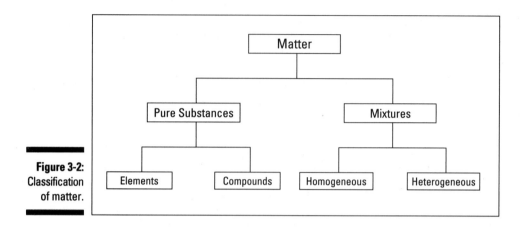

Figure 3-2:
Classification
of matter.

Keeping it simple with pure substances

A *pure substance* has a definite and constant composition or make-up — like salt or sugar. A pure substance can be either an element or a compound, but the composition of a pure substance doesn't vary.

Elementary, my dear reader

An *element* is composed of a single kind of atom. An *atom* is the smallest particle of an element that still has all the properties of the element. Here's an example: Gold is an element. If you slice and slice a chunk of gold until only one tiny particle is left that can't be chopped any more without losing the properties that make gold *gold,* then you've got an atom.

The atoms in an element all have the same number of protons. *Protons* are subatomic particles — particles of an atom. Subatomic particles come in three major kinds, which Chapter 4 covers in great, gory detail.

The important thing to remember right now is that elements are the building blocks of matter. And they're represented in a strange table you may have seen at one time or another — the periodic table. (If you haven't seen such a table before, it's just a list of elements. Chapter 5 contains one if you want to take a peek.)

Compounding the problem

A *compound* is composed of two or more elements in a specific ratio. For example, water (H_2O) is a compound made up of two elements, hydrogen (H) and oxygen (O). These elements are combined in a very specific way — in a ratio of two hydrogen atoms to one oxygen atom (hence H_2O). A lot of compounds contain hydrogen and oxygen, but only one has that special

2-to-1 ratio we call water. Even though water is made up of hydrogen and oxygen, the compound water doesn't have the physical and chemical properties of hydrogen or oxygen — water's properties are unique, entirely different from the two elements of which it's composed.

Chemists can't easily separate the components of a compound: They have to resort to some type of chemical reaction.

Throwing mixtures into the mix

Mixtures are physical combinations of pure substances (elements and/or compounds) that have no definite or constant composition — the composition of a mixture varies according to who prepares the mixture. Suppose I asked two people to prepare me a margarita (a delightful mixture). Unless these two people used exactly the same recipe, these mixtures would vary somewhat in their relative amounts of tequila, triple sec, and so on. They would have produced two slightly different mixtures. However, each component of a mixture (that is, each pure substance that makes up the mixture — in the drink example, each *ingredient*) retains its own set of physical and chemical characteristics. Because of this, it's relatively easy to separate the various substances in a mixture.

Although chemists have a difficult time separating compounds into their specific elements, the different parts of a mixture can be easily separated by physical means, such as filtration. For example, suppose you have a mixture of salt and sand, and you want to purify the sand by removing the salt. You can do this by adding water, dissolving the salt, and then filtering the mixture. You then end up with pure sand.

Mixtures can come in two different forms:

- ✔ **Homogeneous mixtures:** Sometimes called *solutions,* this type of mixture is relatively uniform in composition; every portion of the mixture is like every other portion. If you dissolve sugar in water and mix it really well, your mixture is basically the same no matter where you sample it.

- ✔ **Heterogeneous mixtures:** If you put some sugar in a jar, add some sand, and then give the jar a couple of shakes, your mixture doesn't have the same composition throughout the jar. Because the sand is heavier, there's probably more sand at the bottom of the jar and more sugar at the top. In this case you have a *heterogeneous mixture,* a mixture whose composition varies from position to position within the sample.

Nice Properties You've Got There

When chemists study chemical substances, they examine two types of properties:

- ✔ **Chemical properties:** These properties enable a substance to change into a brand-new substance, and they describe how a substance reacts with other substances. Does a substance change into something completely new when water is added — like how sodium metal changes to sodium hydroxide? Does it burn in air?

- ✔ **Physical properties:** These properties describe the physical characteristics of a substance. The color, luster, hardness, and so on of a substance are its physical properties, and so is its ability to conduct electricity.

Some physical properties are *extensive properties,* properties that depend on the amount of matter present. Mass and volume are extensive properties. A large chunk of gold has a larger mass and volume than a smaller chunk. *Intensive properties,* however, don't depend on the amount of matter present. Hardness is an intensive property. A large chunk of gold, for example, has the same hardness as a small chunk of gold. The mass and volume of these two chunks are different (extensive properties), but the hardness is the same. Intensive properties are especially useful to chemists because they can use intensive properties to identify a substance.

Identifying substances by density

Density is one of the most useful intensive properties of a substance, enabling chemists to more easily identify substances. For example, knowing the differences between the density of quartz and diamond allows a jeweler to check out that engagement ring quickly and easily. *Density (d)* is the ratio of the mass *(m)* to volume *(v)* of a substance. Mathematically, it looks like this:

$$d = \frac{m}{v}$$

Usually, mass is described in grams (g) and volume in milliliters (mL), so density is g/mL. Because the volumes of liquids vary somewhat with temperature, chemists also usually specify the temperature at which a density measurement is made. Most reference books report densities at 20 degrees Celsius, because it's close to room temperature and easy to measure without a lot of heating or cooling. The density of water at 20 degrees, for example, is 1 g/mL.

Another term you may sometimes hear is *specific gravity (sg)*, which is the ratio of the density of a substance to the density of water at the same temperature. Specific gravity is just another way for you to get around the problem of volumes of liquids varying with the temperature. Specific gravity is used with urinalysis in hospitals and to describe automobile battery fluid in auto repair shops. Note that specific gravity has no units of measure associated with it, because the units g/mL appear in both the numerator and the denominator, canceling each other out. In most cases, the density and specific gravity are almost the same, so it's common to simply use the density.

You may sometimes see density reported as g/cm^3 or g/cc. These units are the same as g/mL. A cube measuring 1 centimeter on each edge (written as $1~cm^3$) has a volume of 1 milliliter (1 mL). Because $1~mL = 1~cm^3$, g/mL and g/cm^3 are interchangeable. And because a cubic centimeter (cm^3) is commonly abbreviated cc, g/cc also means the same thing. (You hear cc a lot in the medical profession. When you receive a 10 cc injection, you're getting 10 milliliters of liquid. That's a lot. You better believe I'm running the other way when I see a nurse coming with a 10 cc shot!)

How dense are you? Measuring density

Calculating density is pretty straightforward. You measure the mass of an object by using a balance or scale, determine the object's volume, and then divide the mass by the volume.

Determining the volume of liquids is easy, but solids can be tricky. If the object is a regular solid, like a cube, you can measure its three dimensions and calculate the volume by multiplying the length by the width by the height (volume = $l \times w \times h$). But if the object is an irregular solid, like a rock, determining the volume is more difficult. With irregular solids, you can measure the volume by using something called Archimedes' principle.

Archimedes' principle states that the volume of a solid is equal to the volume of water it displaces. The Greek mathematician Archimedes discovered this concept in the third century BC, and finding an object's density is greatly simplified by using it. Say that you want to measure the volume of a small rock in order to determine its density. First, put some water into a graduated cylinder with markings for every mL and read the volume. (The example in Figure 3-3 shows 25 mL.) Next, put the rock in, making sure that it's totally submerged, and read the volume again (29 mL in Figure 3-3). The difference in volume (4 mL) is the volume of the rock.

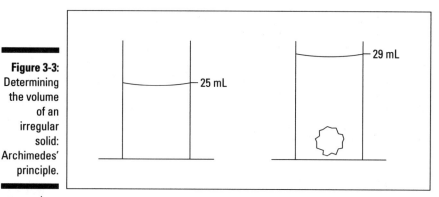

Figure 3-3:
Determining
the volume
of an
irregular
solid:
Archimedes'
principle.

Anything with a density lower than water floats when put into water, and anything with a density greater than 1 g/mL sinks.

For your pondering pleasure, Table 3-1 lists the density of some common materials.

Table 3-1	Densities of Typical Solids and Liquids
Substance	*Density (In g/mL)*
Gasoline	0.68
Ice	0.92
Water	1.00
Table salt	2.16
Iron	7.86
Lead	11.38
Mercury	13.55
Gold	19.3

Note that gold has a pretty high density. One of those gold bars that are stored in Fort Knox weighs over 30 pounds. Remember that fact when you see those burglars on TV throwing a bunch of gold bars into a bag, throwing it over their shoulders, and carrying them away.

If you know the density of a substance and either its mass or volume, you can calculate the other.

Suppose you have a 25.0 mL sample of mercury (d = 13.55 g/mL). What would be the mass of that sample of mercury?

To solve, follow these easy steps:

1. **Start with the density formula.**

$$d = \frac{m}{v}$$

2. **Switch it around so you're solving for the mass.**

$$m = dv$$

3. **Put in your density and volume and solve for the mass (m):**

$$m = \frac{13.55 \text{ g}}{1 \text{ mL}} \times \frac{25.0 \text{ mL}}{1} = 339 \text{ g}$$

That's about ¾ of a pound!

Keeping the World in Motion: Energy

Matter is one of two components of the universe. Energy is the other. *Energy* is the ability to do work. And if you're like I am, at about 5 p.m. your ability to do work — and your energy level — are pretty low.

Energy can take several forms — such as heat energy, light energy, electrical energy, and mechanical energy. But two general categories of energy are especially important to chemists — kinetic energy and potential energy.

Moving right along: Kinetic energy

Kinetic energy is energy of motion. A baseball flying through the air toward a batter has a large amount of kinetic energy (as I'm sure anyone who's ever been hit with a baseball will agree!). Collisions between moving particles and the subsequent transfer of kinetic energy cause chemical reactions to occur. Chemists sometimes study moving particles, especially gases, because understanding the kinetic energy of these particles helps determine whether a particular reaction may take place.

The kinetic energy of moving particles can be transferred from one particle to another. Have you ever shot pool? You transfer kinetic energy from your moving pool stick to the cue ball to (hopefully) the ball you're aiming at.

Kinetic energy can be converted into other types of energy. In a hydroelectric dam, the kinetic energy of the falling water is converted into electrical energy. In fact, a scientific law — *the law of conservation of energy* — states that in ordinary chemical reactions (or physical processes), energy is neither created nor destroyed but can be converted from one form to another. (This law doesn't hold in nuclear reactions, though. Chapter 20 tells you why.)

Sitting pretty: Potential energy

Suppose you take a football and throw it up into a tree where it gets stuck. You gave that ball kinetic energy — energy in motion — when you threw it. But where's that energy now? It's been converted into the other major category of energy — potential energy.

Potential energy is stored energy. Objects may have potential energy stored in terms of their position. That football up in the tree has potential energy due to its height. If the ball were to fall, that potential energy would be converted to kinetic energy. (Watch out!)

Potential energy due to position isn't the only type of potential energy. In fact, chemists really aren't all that interested in potential energy due to position. Chemists are far more interested in the energy stored (potential energy) in *chemical bonds,* which are the forces that hold atoms together in compounds.

It takes a lot of energy to run a human body. What if you couldn't store the energy you extract from food? You'd have to eat all the time just to keep your body going. (My wife claims I eat all the time, anyway!) But humans can store energy in terms of chemical bonds. And then later, when we need that energy, our bodies can break those bonds and release it.

The same is true of the fuels we commonly use to heat our homes and run our automobiles. Energy is stored in these fuels — gasoline, for example — and is released when chemical reactions take place.

Measuring Energy

Measuring kinetic energy is fairly easy. You can do it with a relatively simple instrument — a thermometer. Meanwhile, measuring potential energy can be a difficult task. The potential energy of a ball stuck up in a tree is related to the mass of the ball and its height above the ground. The potential energy contained in chemical bonds is related to the type of bond and the number of bonds that can potentially break. Because measuring potential energy is a tad bit advanced for someone taking a Chem I class, I devote the following sections on how to measure kinetic energy.

Taking a look at temperature

When you measure, say, the air temperature in your backyard, you're really measuring the average kinetic energy (the energy of motion) of the gas particles in your backyard. The faster those particles are moving, the higher the temperature is.

All the particles aren't moving at the same speed. Some are going very fast, and some are going relatively slow, but most are moving at a speed between the two extremes. The temperature reading from your thermometer is related to the *average* kinetic energy of the particles.

You probably use the Fahrenheit scale to measure temperatures, but most scientists and chemists use either the Celsius (°C) or Kelvin (K) temperature scale. (The degree symbol isn't used with K.) Figure 3-4 compares the three temperature scales using the freezing point and boiling point of water as reference points.

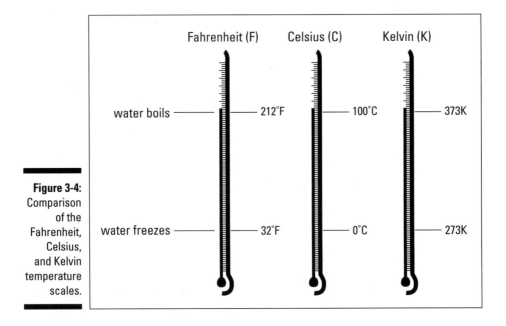

Figure 3-4:
Comparison of the Fahrenheit, Celsius, and Kelvin temperature scales.

As you can see from Figure 3-4, water boils at 100 degrees Celsius (373 Kelvin) and freezes at 0 degrees Celsius (273 Kelvin). To get the Kelvin temperature, you take the Celsius temperature and add 273. Mathematically, it looks like this:

$$K = °C + 273$$

You may want to know how to convert from Fahrenheit to Celsius and vice versa (because most people in the United States still think in degrees Fahrenheit). Here are the equations you need:

$$°C = \frac{5}{9}\left(°F - 32\right)$$

Be sure to subtract 32 from your Fahrenheit temperature before multiplying by ⅝.

$$°F = \frac{9}{5}\left(°C\right) + 32$$

Be sure to multiply your Celsius temperature by ⅝ and *then* add 32.

Go ahead — try these equations out by confirming that the normal body temperature of 98.6 degrees Fahrenheit equals 37 degrees Celsius.

Most of the time in this book, I use the Celsius scale. But when I describe the behavior of gases, I use the Kelvin scale.

Feeling the heat

To a chemist, heat is not the same as temperature. When you measure the temperature of something, you're measuring how hot or cold something is compared to something else. When I describe gases in Chapter 6, I tell you that the temperature of a gas is a measure of the average kinetic energy of the individual particles. *Heat,* on the other hand, is the thermal energy that flows between two objects that are in contact if those objects are at different temperatures. I talk more about heat in Chapter 10.

The unit of heat in the SI system is the *joule (J).* Many chemists also still use the metric unit of heat, the *calorie (cal).* Here's the relationship between the two:

> 1 calorie = 4.184 joule

The calorie is a fairly small amount of heat — the amount it takes to raise the temperature of 1 gram of water 1 degree Celsius. I often use the *kilocalorie (kcal),* which is 1,000 calories, as a convenient unit of heat. If you burn a large kitchen match completely, it produces about 1 kilocalorie (1,000 cal) of heat.

Well, it's definitely not a gas. . . .

I'm sure by now you're familiar with the three states of matter: solid, liquid, and gas. However, sometimes a substance can be difficult to classify into one of these three states. The classic example is cornstarch putty. To see what makes this substance hard to classify, get a pie pan and place a couple cups of cornstarch in it. Using your hands, slowly mix in small amounts of water. Sooner or later, you will reach the point where a non-Newtonian substance has formed. (You can tell you're at this point if you slap the substance hard and it doesn't splatter all over you.)

Pick some up putty in your hand and squeeze it. If you apply pressure rapidly, it turns to a solid. If you release that pressure, the substance flows like a liquid. If you apply pressure slowly, it retains the characteristics of a liquid. Like Silly Putty, your homemade non-Newtonian substance resists getting pigeonholed as a liquid or a solid.

Chapter 4

Something Smaller Than an Atom? Atomic Structure

I remember learning about atoms as a child in school. My teachers called them building blocks, and, in fact, we used blocks to represent atoms. I also remember being told that atoms were so small that nobody would ever see one. Imagine my surprise years later when the first pictures of atoms appeared. They weren't very detailed, but they did make me stop and think how far science had come. I am still amazed when I see pictures of atoms.

In this chapter, I tell you about atoms, the fundamental building blocks of the universe. I cover the three basic particles of an atom — protons, neutrons, and electrons — and show you where they're located. And I devote a slew of pages to electrons themselves, because chemical reactions (where a lot of chemistry comes into play) depend on the loss, gain, or sharing of them.

Taking an Up-Close Look at the Atom: Subatomic Particles

The *atom* is the smallest part of matter that represents a particular element. For quite a while, the atom was thought to be the smallest part of matter that could exist. But in the latter part of the 19th century and early part of the 20th, scientists discovered that atoms are composed of certain subatomic particles and that, no matter what the element, the same *subatomic particles* make up the atom. The number of the various subatomic particles is the only thing that varies.

Scientists now recognize that there are many subatomic particles (this really makes physicists salivate). But in order to be successful in chemistry, you really only need to be concerned with the three major subatomic particles:

- **Protons:** The subatomic particle found in the atom's dense central core that has a positive charge

- **Neutrons:** The subatomic particle found in the atom's dense central core that has no charge

- **Electrons:** The subatomic particle found outside the atom's dense central core that has a negative charge

Table 4-1 summarizes the characteristics of these three subatomic particles.

Table 4-1		The Three Major Subatomic Particles			
Name	*Symbol*	*Charge*	*Mass (g)*	*Mass (amu)*	*Location*
Proton	p^+	+1	1.673×10^{-24}	1	Nucleus
Neutron	n^o	0	1.675×10^{-24}	1	Nucleus
Electron	e^-	−1	9.109×10^{-28}	0.0005	Outside nucleus

In Table 4-1, the masses of the subatomic particles are listed in two ways: grams and *amu,* which stands for *atomic mass units.* Expressing mass in amu is much easier than using the gram equivalent.

Atomic mass units are based on something called the carbon-12 scale, a world-wide standard that's been adopted for atomic weights. By international agreement, a carbon atom that contains 6 protons and 6 neutrons has an atomic weight of exactly 12 amu, so 1 amu is $\frac{1}{12}$ of this carbon atom. (What do carbon atoms and the number 12 have to do with anything? Just trust me.) Because the mass in grams of protons and neutrons are almost exactly the same, both protons and neutrons are said to have a mass of 1 amu. Notice that the mass of an electron is much smaller than that of either a proton or a neutron. It takes almost 2,000 electrons to equal the mass of a single proton.

Table 4-1 also shows the electrical charge associated with each subatomic particle. Matter can be electrically charged in one of two ways: positively or negatively. The proton carries one unit of positive charge, the electron carries one unit of negative charge, and the neutron has no charge — it's neutral.

Scientists have discovered through observation that objects with like charges, whether positive or negative, repel each other, and objects with unlike charges attract each other.

The atom itself has no charge. It's neutral. (Well, actually, later in this chapter, and further in Chapter 13, I explain that certain atoms can gain or lose electrons and acquire a charge. Atoms that gain a charge, either positive or negative, are called *ions*.) So how can an atom be neutral if it contains positively charged protons and negatively charged electrons? Ah, good question. The answer is that it has an *equal* number of protons and electrons, so the equal positive and negative charges cancel each other out.

The last column in Table 4-1 lists the locations of the three subatomic particles. Protons and neutrons are located in the *nucleus,* a dense central core in the middle of the atom, while the electrons are located outside the nucleus (see "Locating the Electrons in an Atom" later in this chapter).

Taking Center Stage: The Nucleus

In 1911 Ernest Rutherford discovered that atoms have a *nucleus* — a center — containing protons. Scientists later discovered that the nucleus also houses the neutron.

The nucleus is very, very small and very, very dense when compared to the rest of the atom. Typically, atoms have diameters that measure around 10^{-10} meters. (That's small!) Nuclei are around 10^{-15} meters in diameter. (That's *really* small!) For example, if the Superdome in New Orleans represented a hydrogen atom, the nucleus would be about the size of a pea.

The protons of an atom are all crammed together inside the nucleus. Now some of you may be thinking, "Okay, each proton carries a positive charge, and like charges repel each other. So if all the protons are repelling each other, why doesn't the nucleus simply fly apart?" It's *The Force,* Luke. Forces in the nucleus counteract this repulsion and hold the nucleus together. (Physicists call these forces *nuclear glue.* But sometimes this "glue" isn't strong enough, and the nucleus does break apart. This process is called *radioactivity.*)

Not only is the nucleus very small, but it also contains most of the mass of the atom. In fact, for all practical purposes, the mass of the atom is the sum of the masses of the protons and neutrons. (I tend to ignore the minute mass of the electrons unless I'm doing very, very precise calculations.)

The sum of the number of protons plus the number of neutrons in an atom is called the *mass number.* And the number of protons in a particular atom is given a special name, the *atomic number.* Chemists commonly use the symbolization shown in Figure 4-1 to represent these things for a particular element.

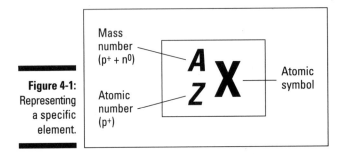

Figure 4-1:
Representing
a specific
element.

As shown in Figure 4-1, chemists use the placeholder X to represent the chemical symbol. You can find an element's chemical symbol on the periodic table or in a list of elements (see Table 4-2 for a list of elements). (Not all the known elements are included in the table, just all the ones you'll be using in your chemistry class.) The placeholder Z represents the atomic number — the number of protons in the nucleus. And A represents the mass number, the sum of the number of protons plus neutrons. The mass number is listed in amu.

Table 4-2

The Elements

Name	Symbol	Atomic Number	Mass Number	Name	Symbol	Atomic Number	Mass Number
Actinium	Ac	89	227.028	Cerium	Ce	58	140.115
Aluminum	Al	13	26.982	Cesium	Cs	55	132.905
Americium	Am	95	243	Chlorine	Cl	17	35.453
Antimony	Sb	51	121.76	Chromium	Cr	24	51.996
Argon	Ar	18	39.948	Cobalt	Co	27	58.933
Arsenic	As	33	74.922	Copper	Cu	29	63.546
Astatine	At	85	210	Curium	Cm	96	247
Barium	Ba	56	137.327	Dubnium	Db	105	262
Berkelium	Bk	97	247	Dysprosium	Dy	66	162.5
Beryllium	Be	4	9.012	Einsteinium	Es	99	252
Bismuth	Bi	83	208.980	Erbium	Er	68	167.26
Bohrium	Bh	107	262	Europium	Eu	63	151.964
Boron	B	5	10.811	Fermium	Fm	100	257
Bromine	Br	35	79.904	Fluorine	F	9	18.998
Cadmium	Cd	48	112.411	Francium	Fr	87	223
Calcium	Ca	20	40.078	Gadolinium	Gd	64	157.25
Californium	Cf	98	251	Gallium	Ga	31	69.723
Carbon	C	6	12.011	Germanium	Ge	32	72.61

(continued)

Table 4-2 (continued)

Name	Symbol	Atomic Number	Mass Number	Name	Symbol	Atomic Number	Mass Number
Gold	Au	79	196.967	Mendelevium	Md	101	258
Hafnium	Hf	72	178.49	Mercury	Hg	80	200.59
Hassium	Hs	108	265	Molybdenum	Mo	42	95.94
Helium	He	2	4.003	Neodymium	Nd	60	144.24
Holmium	Ho	67	164.93	Neon	Ne	10	20.180
Hydrogen	H	1	1.0079	Neptunium	Np	93	237.048
Indium	In	49	114.82	Nickel	Ni	28	58.69
Iodine	I	53	126.905	Niobium	Nb	41	92.906
Iridium	Ir	77	192.22	Nitrogen	N	7	14.007
Iron	Fe	26	55.845	Nobelium	No	102	259
Krypton	Kr	36	83.8	Osmium	Os	76	190.23
Lanthanum	La	57	138.906	Oxygen	O	8	15.999
Lawrencium	Lr	103	262	Palladium	Pd	46	106.42
Lead	Pb	82	207.2	Phosphorus	P	15	30.974
Lithium	Li	3	6.941	Platinum	Pt	78	195.08
Lutetium	Lu	71	174.967	Plutonium	Pu	94	244
Magnesium	Mg	12	24.305	Polonium	Po	84	209
Manganese	Mn	25	54.938	Potassium	K	19	39.098
Meitnerium	Mt	109	266	Praseodymium	Pr	59	140.908

Name	Symbol	Atomic Number	Mass Number	Name	Symbol	Atomic Number	Mass Number
Promethium	Pm	61	145	Tantalum	Ta	73	180.948
Protactinium	Pa	91	231.036	Technetium	Tc	43	98
Radium	Ra	88	226.025	Tellurium	Te	52	127.60
Radon	Rn	86	222	Terbium	Tb	65	158.925
Rhenium	Re	75	186.207	Thallium	Tl	81	204.383
Rhodium	Rh	45	102.906	Thorium	Th	90	232.038
Rubidium	Rb	37	85.468	Thulium	Tm	69	168.934
Ruthenium	Ru	44	101.07	Tin	Sn	50	118.71
Rutherfordium	Rf	104	261	Titanium	Ti	22	47.88
Samarium	Sm	62	150.36	Tungsten	W	74	183.84
Scandium	Sc	21	44.956	Uranium	U	92	238.029
Seaborgium	Sg	106	263	Vanadium	V	23	50.942
Selenium	Se	34	78.96	Xenon	Xe	54	131.29
Silicon	Si	14	28.086	Ytterbium	Yb	70	173.04
Silver	Ag	47	107.868	Yttrium	Y	39	88.906
Sodium	Na	11	22.990	Zinc	Zn	30	65.39
Strontium	Sr	38	87.62	Zirconvium	Zr	40	91.224
Sulfur	S	16	32.066				

Suppose you want to represent uranium. The mass number of a particular element isn't shown on the periodic table. What is shown is the average *atomic mass,* or *atomic weight,* for all forms of that particular element, taking into account the amounts of each found in nature. So you can represent uranium as shown in Figure 4-2.

Figure 4-2:
Representing
uranium.

$^{238}_{92}U$

You know that uranium has an atomic number of 92 (number of protons) and a mass number of 238 (protons plus neutrons). So if you want to know the number of neutrons in uranium, all you have to do is subtract the atomic number (92 protons) from the mass number (238 protons plus neutrons). The resulting number shows that uranium has 146 neutrons.

But how many electrons does uranium have? Because the atom is neutral (it has no electrical charge), an equal number of positive and negative charges — protons and electrons — must be inside it. So each uranium atom has 92 electrons.

Locating the Electrons in an Atom

Early models of the atom had electrons spinning around the nucleus in a random fashion or like planets circling the sun. But as scientists learned more about the atom, they found that this representation probably wasn't accurate. Today, two models of atomic structure are used: the Bohr model and the quantum mechanical model. The Bohr model is simple and relatively easy to understand; the quantum mechanical model is based on mathematics and is more difficult to understand. Both, though, are helpful in understanding the atom, so I explain each in the following sections (without resorting to a lot of math).

A model is useful because it helps you understand what's observed in nature. It's not unusual to have more than one model represent and help people understand a particular topic.

The Bohr model — it's really not boring

Have you ever bought color crystals for your fireplace to make flames of different colors? Or have you ever watched fireworks and wondered where the colors came from?

Color comes from different elements. If you sprinkle table salt — or any salt containing sodium — on a fire, you get a yellow color. Salts that contain copper give a greenish-blue flame. And if you look at the flames through a *spectroscope,* an instrument that uses a prism to break up light into its various components, you see a number of lines of various colors. Those distinct lines of color make up a *line spectrum.*

Niels Bohr, a Danish scientist, explained this line spectrum while developing a model for the atom. The Bohr model shows that the electrons in atoms are in orbits of differing energy around the nucleus. Bohr used the term *energy levels* (or *shells*) to describe these orbits of differing energy. And he said that the energy of an electron is *quantized,* meaning electrons can be in one energy level or another but not in between.

The energy level an electron normally occupies is called its *ground state.* But it can move to a higher-energy, less-stable level, or shell, by absorbing energy. This higher-energy, less-stable state is called the electron's *excited state.*

After it's done being excited, the electron can return to its original ground state by releasing the energy it has absorbed (see Figure 4-3). And here's where the line spectrum explanation comes in. Sometimes the energy released by electrons occupies the portion of the *electromagnetic spectrum* (the range of wavelengths of energy) that humans detect as visible light. Slight variations in the amount of the energy are seen as light of different colors.

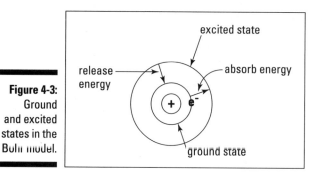

Figure 4-3:
Ground and excited states in the Bohr model.

Bohr found that the closer an electron is to the nucleus, the less energy it possesses, but the farther away it is, the more energy it possesses. So Bohr numbered the electron's energy levels. The higher the energy-level number, the farther away the electron is from the nucleus — and the higher the energy.

Bohr also found that the various energy levels can hold differing numbers of electrons: Energy level 1 may hold up to two electrons, energy level 2 may hold up to eight electrons, and so on.

The Bohr model worked well for very simple atoms such as hydrogen (which has one electron) but not for more complex atoms. Although the Bohr model is still used today, especially in elementary textbooks, a more sophisticated (and complex) model — the quantum mechanical model — is used much more frequently.

Quantum mechanical model

The simple Bohr model was unable to explain observations made on complex atoms, so a more complex, highly mathematical model of atomic structure was developed — the quantum mechanical model.

This model is based on *quantum theory,* which says matter also has properties associated with waves. According to quantum theory, knowing both the exact position and *momentum* (speed and direction) of an electron at the same time is impossible. This fact is known as the *uncertainty principle.* So scientists had to replace Bohr's orbits with *orbitals* (sometimes called *electron clouds*), volumes of space in which an electron is *likely* to be. In other words, certainty was replaced with probability.

The quantum mechanical model of the atom uses complex shapes of orbitals rather than Bohr's simple circular orbits. Without resorting to a lot of math (you're welcome), this section shows you some aspects of this newest model of the atom.

Four numbers, called *quantum numbers,* were introduced to describe the characteristics of electrons and their orbitals. You'll notice that they were named by totally top-rate techno-geeks:

- ✔ Principal quantum number n
- ✔ Angular momentum quantum number l
- ✔ Magnetic quantum number m_l
- ✔ Spin quantum number m_s

Table 4-3 summarizes the four quantum numbers. When the numbers are all put together, theoretical chemists have a pretty good description of the characteristics of a particular electron.

Table 4-3	Summary of the Quantum Numbers		
Name	*Symbol*	*Description*	*Allowed Values*
Principal	n	Orbital energy	Positive integers (1, 2, 3, and so on)
Angular momentum	l	Orbital shape	Integers from 0 to $n-1$
Magnetic	m_l	Orientation	Integers from $-l$ to 0 to $+l$
Spin	m_s	Electron spin	$+\frac{1}{2}$ or $-\frac{1}{2}$

The principal quantum number n

The principal quantum number n describes the average distance of the orbital from the nucleus — and the energy of the electron in an atom. It's really about the same as Bohr's energy-level numbers. It can have positive integer (whole number) values: 1, 2, 3, 4, and so on. The larger the value of n, the higher the energy and the larger the orbital. Chemists sometimes call the orbitals *electron shells*.

The angular momentum quantum number l

The angular momentum quantum number l describes the shape of the orbital, and the shape is limited by the principal quantum number n: The angular momentum quantum number l can have positive integer values from 0 to $n-1$. For example, if the n value is 3, three values are allowed for l: 0, 1, and 2.

The value of l defines the shape of the orbital, and the value of n defines the size. Orbitals that have the same value of n but different values of l are called *subshells*. These subshells are given different letters to help chemists distinguish them from each other. Table 4-4 shows the letters corresponding to the different values of l.

Table 4-4	Letter Designation of the Subshells
Value of l (subshell)	*Letter*
0	s
1	p
2	d
3	f
4	g

When chemists describe one particular subshell in an atom, they can use both the n value and the subshell letter — 2p, 3d, and so on. Normally, a subshell value of 4 is the largest needed to describe a particular subshell. If chemists ever need a larger value, they can create subshell numbers and letters. Figure 4-4 shows the shapes of the s, p, and d orbitals.

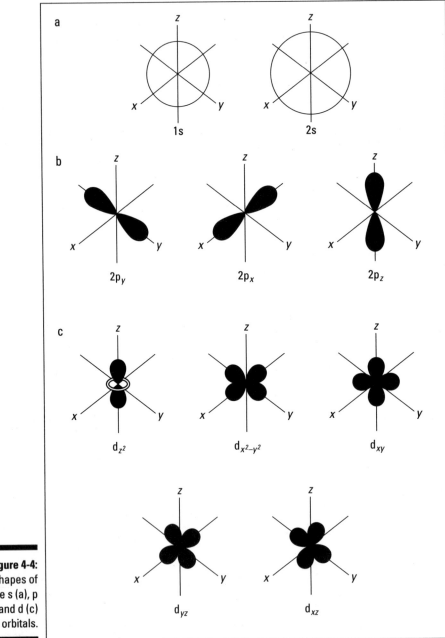

Figure 4-4:
Shapes of
the s (a), p
(b), and d (c)
orbitals.

Figure 4-4a has two s orbitals — one for energy level 1 (1s) and the other for energy level 2 (2s). S orbitals are spherical with the nucleus at the center. Notice that the 2s orbital is larger in diameter than the 1s orbital. In large atoms the 1s orbital is nestled inside the 2s, just like the 2p is nestled inside the 3p.

Figure 4-4b shows the shapes of the p orbitals, and Figure 4-4c shows the shapes of the d orbitals. Notice that the shapes get progressively more complex.

The magnetic quantum number m_l

The magnetic quantum number m_l describes how the various orbitals are oriented in space. The value of m_l depends on the value of l. The values allowed are integers from $-l$ to 0 to $+l$. For example, if the value of $l = 1$ (p orbital — refer to Table 4-4), you can write three values for m_l: -1, 0, and $+1$. This means that there are three different p subshells for a particular orbital. The subshells have the same energy but different orientations in space.

Figure 4-4b shows how the p orbitals are oriented in space. Notice that the three p orbitals correspond to m_l values of -1, 0, and $+1$, oriented along the x, y, and z axes.

The spin quantum number m_s

The fourth and final (I know you're glad — techie stuff, eh?) quantum number is the spin quantum number m_s. This one you can think of as describing the direction the electron is spinning in a magnetic field — either clockwise or counterclockwise. Only two values are allowed for m_s: $+\frac{1}{2}$ or $-\frac{1}{2}$. Each subshell can have only two electrons, one with a spin of $+\frac{1}{2}$ and another with a spin of $-\frac{1}{2}$.

Put all the numbers together and whaddya get? (A pretty table)

I know. Quantum number stuff makes science nerds drool and normal people yawn. But, hey, sometime if the TV's on the blink and you've got some time to kill, take a peek at Table 4-5. You can check out the quantum numbers for each electron in the first two energy levels (oh boy, oh boy, oh boy).

Table 4-5   Quantum Numbers for the First Two Energy Levels

n	l	Subshell Notation	m_l	m_s
1	0	1s	0	$+\frac{1}{2}, -\frac{1}{2}$
2	0	2s	0	$+\frac{1}{2}, -\frac{1}{2}$
	1	2p	-1	$+\frac{1}{2}, -\frac{1}{2}$
			0	$+\frac{1}{2}, -\frac{1}{2}$
			$+1$	$+\frac{1}{2}, -\frac{1}{2}$

Table 4-5 shows that energy level 1 ($n = 1$) has only an s orbital. It has no p orbital because an l value of 1 (p orbital) is not allowed. And notice that the 1s orbital can have only two electrons (m_s of +½ and –½). In fact, a maximum of only two electrons can be in any s orbital, whether it's 1s or 5s.

When you move from energy level 1 to energy level 2 ($n = 2$), both s and p orbitals can be present. If you write out the quantum numbers for energy level 3, you see s, p, and d orbitals. Each time you move higher in a major energy level, you add another orbital type.

Notice also that the 2p orbital has three subshells (m_l) (refer to Figure 4-4b) and that each holds a maximum of two electrons. The three 2p subshells can hold a maximum of six electrons.

The major energy levels have an energy difference (energy level 2 is higher in energy than energy level 1), but the energies of the different orbitals within an energy level also have differences. At energy level 2, both s and p orbitals are present. But the 2s is slightly lower in energy than the 2p. The three subshells of the 2p orbital have the same energy. Likewise, the five subshells of the d orbitals (refer to Figure 4-4c) have the same energy.

Configuring Electrons (Bed Check for Electrons)

Chemists find quantum numbers useful when they're looking at chemical reactions and bonding (and those are things many chemists like to study). But they find two other representations for electrons *more* useful and easier to work with:

- ✔ Energy-level diagrams
- ✔ Electron configurations

Chemists (and chemistry students) use both of these things to represent which energy level, subshell, and orbital are occupied by electrons in any particular atom. Chemists (and chemistry students) use this information to predict what type of bonding will occur with a particular element and show exactly which electrons are being used. These representations are also useful in showing why certain elements behave in similar ways.

In this section, I show you how to use an energy-level diagram and write electron configurations.

Examining the energy-level diagram

Figure 4-5 is a blank energy-level diagram you can use to depict electrons for any particular atom. Not all the known orbitals and subshells are shown. But with this diagram, you should be able to do just about anything you need to. (If you don't have a clue what orbitals, subshells, or all those numbers and letters in the figure have to do with the price of beans, check out the "Quantum mechanical model" section, earlier in this chapter. Fun read, lemme tell ya.)

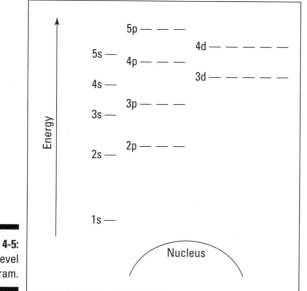

Figure 4-5:
Energy-level
diagram.

I represent orbitals with dashes in which you can place a maximum of two electrons. The 1s orbital is closest to the nucleus, and it has the lowest energy. It's also the only orbital in energy level 1 (refer to Table 4-5). At energy level 2 are both s and p orbitals, with the 2s having lower energy than the 2p. The three 2p subshells are represented by three dashes of the same energy. Energy levels 3, 4, and 5 are also shown. Notice that the 4s has lower energy than the 3d. This is an exception to what you may have thought, but it's what's observed in nature. Go figure. Speaking of which, Figure 4-6 shows the *aufbau principle,* a method for remembering the order in which orbitals fill the vacant energy levels.

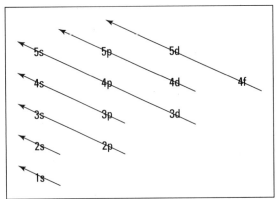

Figure 4-6:
Aufbau fill-
ing chart.

In using the energy-level diagram (Figure 4-5), remember two things:

✔ Electrons fill the lowest vacant energy levels first.

✔ When more than one subshell lies at a particular energy level, such as at the 3p or 4d level (refer to Figure 4-5), only one electron fills each sub-shell until each subshell has one electron. Then electrons start pairing up in each subshell. This rule is named *Hund's rule.*

Suppose you want to draw the energy-level diagram of oxygen. You look on the periodic table or an element list and find that oxygen is atomic number 8. This number means that oxygen has 8 protons in its nucleus and 8 electrons, so you put 8 electrons into your energy-level diagram. You can represent electrons as arrows (see Figure 4-7). Note that if two electrons end up in the same orbital, one arrow faces up and the other faces down. (This arrangement is called *spin pairing.* It corresponds to the $+\frac{1}{2}$ and $-\frac{1}{2}$ of m_s — see "The spin quantum number m_s" section earlier in this chapter.)

The first electron goes into the 1s orbital, filling the lowest energy level first, and the second one spin pairs with the first one. Electrons 3 and 4 spin pair in the next lowest vacant orbital — the 2s. Electron 5 goes into one of the 2p subshells (no, it doesn't matter which one — they all have the same energy), and electrons 6 and 7 go into the other two totally vacant 2p orbitals (following Hund's rule). The last electron spin pairs with one of the electrons in the 2p subshells (again, it doesn't matter which one you pair it with). Figure 4-7 shows the completed energy-level diagram for oxygen.

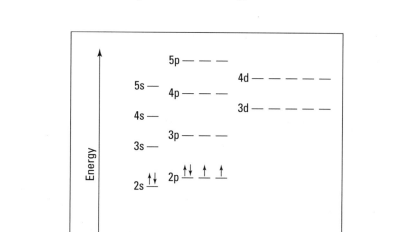

Figure 4-7:
Energy-level
diagram for
oxygen.

Eyeing electron configurations

Energy-level diagrams are useful when you need to figure out chemical reactions and bonding, but they're very bulky to work with. Wouldn't it be nice if a different representation gave just about the same information but in a much more concise, shorthand-notation form? Well, that representation exists. It's called the *electron configuration.*

The electron configuration for oxygen is $1s^22s^22p^4$. Compare that notation with the energy-level diagram for oxygen in Figure 4-7. Doesn't the electron configuration take up a lot less space? You can derive the electron configuration from the energy-level diagram. The first two electrons in oxygen fill the 1s orbital, so you show it as $1s^2$ in the electron configuration. The 1 is the energy level, the s represents the type of orbital, and the superscript 2 represents the number of electrons in that orbital. The next two electrons are in the 2s orbital, so you write $2s^2$. And, finally, you show the 4 electrons in the 2p orbital as $2p^4$. Put it all together, and you get $1s^22s^22p^4$.

Some people use a more expanded form, showing how the individual p_x, p_y, and p_z orbitals are oriented along the x, y, and z axes and the number of electrons in each orbital. (The earlier section "The magnetic quantum number m_l," explains how orbitals are oriented in space.) The expanded form is nice if you're really looking at the finer details, but most of the time you don't need that amount of detail in order to show bonding situations and such, so I don't explain the expanded form here.

The sum of the superscript numbers equals the atomic number, or the number of electrons in the atom.

Here are a couple of electron configurations you can use to check your conversions from energy-level diagrams:

Chlorine (Cl): $1s^22s^22p^63s^23p^5$

Iron (Fe): $1s^22s^22p^63s^23p^64s^23d^6$

Although I've showed you how to use the energy-level diagram to write the electron configuration, with a little practice, you can omit doing the energy-level diagram altogether and simply write the electron configuration by knowing the number of electrons and the orbital filling pattern. Anything to save a little precious time, right?

Living on the edge: Valence electrons

When chemists study chemical reactions, they study the transfer or sharing of electrons. The electrons more loosely held by the nucleus — the electrons in the energy level farthest away from the nucleus — are the ones that are gained, lost, or shared.

Electrons are negatively charged, whereas the nucleus has a positive charge due to the protons. The protons attract and hold the electrons, but the farther away the electrons are, the less the attractive force.

The electrons in the outermost energy level are commonly called *valence electrons*. Chemists really only consider the electrons in the s and p orbitals in the energy level that is currently being filled as valence electrons. In the electron configuration for oxygen, $1s^22s^22p^4$, energy level 1 is filled, and in energy level 2, two electrons are in the 2s orbital and four electrons are in the 2p orbital for a total of 6 valence electrons. Those valence electrons are the ones lost, gained, or shared.

Being able to determine the number of valence electrons in a particular atom gives you a big clue as to how that atom will react. In Chapter 5, which gives an overview of the periodic table, I show you a quick way to determine the number of valence electrons without writing the electron configuration of the atom.

Examining Isotopes and Ions

The atoms in a particular element have an identical number of protons and electrons but can have varying numbers of neutrons. If they have different numbers of neutrons, then the atoms are called *isotopes*. If the neutral atom gains or loses electrons, then the resulting charged atom is called an *ion*. In the following sections, I discuss isotopes and ions.

Isolating the isotope

Hydrogen is a common element here on earth. Hydrogen's atomic number is 1 — its nucleus contains one proton. The hydrogen atom also has one electron. Because it has the same number of protons as electrons, like all other atoms, the hydrogen atom is neutral (the positive and negative charges have canceled each other out).

Most of the hydrogen atoms on Earth contain no neutrons. You can use the symbolization shown in Figure 4-2 to represent hydrogen atoms that don't contain neutrons, as shown in Figure 4-8a.

Figure 4-8: Hydrogen isotopes containing no neutrons (a), one neutron (b), and two neutrons (c).

$$_1^1\text{H}$$ (a) Hydrogen H – 1 $$_1^2\text{H}$$ (b) Deuterium H – 2 $$_1^3\text{H}$$ (c) Tritium H – 3

However, approximately one hydrogen atom out of 6,000 contains a neutron in its nucleus. These atoms are still hydrogen because they have one proton and one electron; they simply have a neutron that most hydrogen atoms lack. These atoms are called isotopes. Figure 4-8b shows an isotope of hydrogen, commonly called *deuterium,* that contains one neutron. Because it contains one proton and one neutron, its mass number is 2.

Hydrogen even has an isotope containing two neutrons. This one's called *tritium,* and it's represented in Figure 4-8c. Only a small amount of tritium occurs naturally on earth, but it can easily be created.

Now take another look at Figure 4-8. Below the name of each symbol is an alternative way of representing isotopes: Write the element symbol, a dash, and then the mass number.

You may be wondering, "If I'm doing a calculation involving the atomic mass of hydrogen, which isotope do I use?" Well, you use an average of all the naturally occurring isotopes of hydrogen. But not a simple average. (You have to take into consideration that there's a *lot* more H-1 than H-2, and you don't even consider H-3, because it's so very rare in nature.) You use a *weighted average,* which takes into consideration the abundances of the naturally occurring isotopes. That's why the atomic mass of hydrogen in Table 4-2 isn't a whole number: It's 1.0079 amu. The number shows that there's a lot more H-1 than H-2.

Silver exists as two isotopes. 51.830 percent of all silver is Ag-107 with a mass of 106.905 amu. The remainder, 48.170 percent, is Ag-109 with a mass of 108.905 amu. What is the weighted average?

To calculate the weighted average, you multiply the percentage (expressed as a decimal) and the mass for each isotope and then add the products together. Following is the calculation for this silver example:

$$(0.51830 \times 106.905 \text{ amu}) + (0.48170 \times 108.905 \text{ amu}) = 107.87 \text{ amu}$$

Many elements have several isotopic forms. You can find out more about them in Chapter 20.

Keeping an eye on ions

Because an atom itself is neutral, throughout this book I say that the number of protons and electrons in atoms are equal. But in some cases an atom can acquire an electrical charge. For example, in the compound sodium chloride — table salt — the sodium atom has a positive charge and the chlorine atom has a negative charge. Atoms (or groups of atoms) in which there are unequal numbers of protons and electrons are called *ions.*

The neutral sodium atom has 11 protons and 11 electrons, which means it has 11 positive charges and 11 negative charges. Overall, the sodium atom is neutral, and it's represented as Na. But the sodium *ion* contains one more positive charge than negative charge, so it's represented as Na$^+$ (the $^+$ represents its net positive electrical charge).

This unequal number of negative and positive charges can occur in one of two ways: An atom can gain a proton (a positive charge) or lose an electron (a negative charge). So which process is more likely to occur? Well, the rough guideline is that gaining or losing electrons is easy but gaining or losing protons is very difficult. So atoms become ions by gaining or losing electrons.

If an ion is formed by the loss of an electron, the ion has more protons than electrons, or more *positive* charges. Those positive ions are called *cations*. You represent the overall positive charge in cations with the little plus sign in the notation (like Na$^+$). If the atom loses two electrons instead of one, the result is still a cation, but it has a stronger positive charge (actually twice as strong as in the case in which only one electron was lost) and is represented with the number of electrons lost *and* a plus sign (like Mg^{2+} for a magnesium cation with two missing electrons, or Al^{3+} for aluminum with three electrons gone).

If an ion is created by gaining an electron, the number of electrons exceeds the number of protons, so the ion acquires a negative charge. Negatively charged ions are called *anions,* and they're represented with a little negative sign ($^-$). If chlorine (Cl) gains an electron, it becomes a chlorine ion because it has unequal numbers of protons and electrons, and as an anion (a negatively charged ion), it's represented as Cl$^-$. (You can get the full scoop on ions, cations, and anions in Chapter 13, if you're interested. This here's just a teaser.)

Just for kicks, here are some extra tidbits about ions for your reading pleasure:

✔ You can write electron configurations and energy-level diagrams for ions. The neutral sodium atom (11 protons) has an electron configuration of $1s^22s^22p^63s^1$. The sodium cation has lost an electron — the valence electron, which is *farthest* away from the nucleus (the 3s electron, in this case). The electron configuration of Na$^+$ is $1s^22s^22p^6$.

✔ If two chemical species have the same electron configuration, they're said to be *isoelectronic*. For example, the electron configuration of the chloride ion (Cl$^-$) is $1s^22s^22p^63s^23p^6$, which is the same electron configuration as the neutral argon atom. Therefore the chloride anion and argon are isoelectronic. Figuring out chemistry requires learning a whole new language, eh?

✔ This section has been discussing *monoatomic* (one-atom) ions. But *polyatomic* (many-atom) ions do exist. The ammonium ion, NH$_4^+$, is a polyatomic ion, or, specifically, a polyatomic cation. The nitrate ion, NO$_3^-$, is also a polyatomic ion, or, specifically, a polyatomic anion.

✔ Ions are commonly found in a class of compounds called *salts,* or *ionic solids.* Salts, when melted or dissolved in water, yield solutions that conduct electricity. A substance that conducts electricity when melted or dissolved in water is called an *electrolyte.* Salts are electrolytes, but as you see in Chapter 11, not all electrolytes are salts. Table salt — sodium chloride — is a good example of an electrolyte.

On the other hand, when table sugar (sucrose) is dissolved in water, it becomes a solution that doesn't conduct electricity. So sucrose is a *nonelectrolyte.* Whether a substance is an electrolyte or a nonelectrolyte gives clues to the type of bonding in the compound. If the substance is an electrolyte, the compound is probably *ionically bonded* (see Chapter 13). If it's a nonelectrolyte, it's probably *covalently bonded* (see Chapter 14).

Chapter 5

The Periodic Table (But No Chairs)

*I*n this chapter, I introduce you to the second most important tool a chemist possesses — the periodic table. (The most important? The beaker and Bunsen burner he or she brews coffee with.)

Chemists are a little lazy, as are most scientists. They like to put things together into groups based on similar properties. This process, called *classification,* makes studying a particular system much easier. Scientists have grouped the elements together in the periodic table so they don't have to learn the properties of individual elements; they can just learn the properties of the various groups. In this chapter, I show you how the elements are arranged in the table and introduce you to some important groups. I also explain how chemists and other scientists go about using the periodic table.

Repeating Patterns of Periodicity

In nature, as well as in inventions of mankind, you may notice some repeating patterns. The seasons repeat their pattern of fall, winter, spring, and summer. The tides repeat their pattern of rising and falling. Tuesday follows Monday, December follows November, and so on. This pattern of repeating order is called *periodicity.*

In the mid-1800s, Dmitri Mendeleev, a Russian chemist, noticed a repeating pattern of chemical properties in the elements that were known at the time. Mendeleev arranged the elements in order of increasing atomic mass (see Chapter 4 for a description of atomic mass) to form something that fairly closely resembles the modern periodic table. He was even able to predict the properties of some of the then-unknown elements. Later, the elements were rearranged in order of increasing *atomic number,* the number of protons in the nucleus of the atom (again, see Chapter 4). Figure 5-1 shows the modern periodic table.

PERIODIC TABLE OF THE ELEMENTS

	1 IA								
1	1 H Hydrogen 1.00797	2 IIA							
2	3 Li Lithium 6.939	4 Be Beryllium 9.0122							
3	11 Na Sodium 22.9898	12 Mg Magnesium 24.312	3 IIIB	4 IVB	5 VB	6 VIB	7 VIIB	8 VIIIB	9 VIIIB
4	19 K Potassium 39.102	20 Ca Calcium 40.08	21 Sc Scandium 44.956	22 Ti Titanium 47.90	23 V Vanadium 50.942	24 Cr Chromium 51.996	25 Mn Manganese 54.9380	26 Fe Iron 55.847	27 Co Cobalt 58.9332
5	37 Rb Rubidium 85.47	38 Sr Strontium 87.62	39 Y Yttrium 88.905	40 Zr Zirconium 91.22	41 Nb Niobium 92.906	42 Mo Molybdenum 95.94	43 Tc Technetium (99)	44 Ru Ruthenium 101.07	45 Rh Rhodium 102.905
6	55 Cs Cesium 132.905	56 Ba Barium 137.34	57 La Lanthanum 138.91	72 Hf Hafnium 179.49	73 Ta Tantalum 180.948	74 W Tungsten 183.85	75 Re Rhenium 186.2	76 Os Osmium 190.2	77 Ir Iridium 192.2
7	87 Fr Francium (223)	88 Ra Radium (226)	89 Ac Actinium (227)	104 Rf Rutherfordium (261)	105 Db Dubnium (262)	106 Sg Seaborgium (266)	107 Bh Bohrium (264)	108 Hs Hassium (269)	109 Mt Meitnerium (268)

Lanthanide Series	58 Ce Cerium 140.12	59 Pr Praseodymium 140.907	60 Nd Neodymium 144.24	61 Pm Promethium (145)	62 Sm Samarium 150.35	63 Eu Europium 151.96
Actinide Series	90 Th Thorium 232.038	91 Pa Protactinium (231)	92 U Uranium 238.03	93 Np Neptunium (237)	94 Pu Plutonium (242)	95 Am Americium (243)

Figure 5-1:
The periodic table.

							18 0
							2 He Helium 4.0026

			13 IIIA	14 IVA	15 VA	16 VIA	17 VIIA	
			5 B Boron 10.811	6 C Carbon 12.01115	7 N Nitrogen 14.0067	8 O Oxygen 15.9994	9 F Fluorine 18.9984	10 Ne Neon 20.183
10 VIIIB	11 IB	12 IIB	13 Al Aluminum 26.9815	14 Si Silicon 28.086	15 P Phosphorus 30.9738	16 S Sulfur 32.064	17 Cl Chlorine 35.453	18 Ar Argon 39.948
28 Ni Nickel 58.71	29 Cu Copper 63.546	30 Zn Zinc 65.37	31 Ga Gallium 69.72	32 Ge Germanium 72.59	33 As Arsenic 74.9216	34 Se Selenium 78.96	35 Br Bromine 79.904	36 Kr Krypton 83.80
46 Pd Palladium 106.4	47 Ag Silver 107.868	48 Cd Cadmium 112.40	49 In Indium 114.82	50 Sn Tin 118.69	51 Sb Antimony 121.75	52 Te Tellurium 127.60	53 I Iodine 126.9044	54 Xe Xenon 131.30
78 Pt Platinum 195.09	79 Au Gold 196.967	80 Hg Mercury 200.59	81 Tl Thallium 204.37	82 Pb Lead 207.19	83 Bi Bismuth 208.980	84 Po Polonium (210)	85 At Astatine (210)	86 Rn Radon (222)
110 Ds Darmstadtium (281)	111 Rg Roentgenium (286)	112 Cn Copernicium (285)	113 Uut Ununtrium (284)	114 Uuq Ununquadium (289)	115 Uup Unupentium (288)	116 Uuh Ununhexium (298)	117 Uus Ununseptium (294)	118 Uuo Ununoctium (294)

64 Gd Gadolinium 157.25	65 Tb Terbium 158.924	66 Dy Dysprosium 162.50	67 Ho Holmium 164.930	68 Er Erbium 167.26	69 Tm Thulium 168.934	70 Yb Ytterbium 173.04	71 Lu Lutetium 174.97
96 Cm Curium (247)	97 Bk Berkelium (247)	98 Cf Californium (251)	99 Es Einsteinium (254)	100 Fm Fermium (257)	101 Md Mendelevium (258)	102 No Nobelium (259)	103 Lr Lawrencium (260)

Chemists can't imagine doing much of anything without having access to the periodic table. Instead of mastering the properties of 118+ elements (more are created almost every year), chemists — and chemistry students — can simply get a firm grasp of the properties of families of elements, thus saving a lot of time and effort. You can find the relationships among elements and figure out the formulas of many different compounds by referring to the periodic table. The table readily provides atomic numbers, mass numbers, and information about the number of valence electrons, outermost s and p electrons.

I remember reading a science-fiction story many years ago about an alien life based on the element silicon. Silicon was a plausible choice for this story because it's in the same family as carbon, the element that's the basis for life on earth. So the periodic table is an absolute necessity for chemists, chemistry students, and science-fiction novelists. Don't leave home without it!

Understanding How Elements Are Arranged in the Periodic Table

Refer to the periodic table in Figure 5-1. The elements are arranged in order of increasing atomic number. The *atomic number* (number of protons) is located right above the element symbol. Under the element symbol is the *atomic mass,* or *atomic weight* (sum of the protons and neutrons). Atomic mass is a *weighted average* of all naturally occurring isotopes. (And if that's Greek to you, just flip to Chapter 4 for tons of fun with atomic mass and isotopes.) Notice also that two rows of elements — Ce–Lu (commonly called the lanthanides) and Th–Lr (the actinides) — have been pulled out of the main body of the periodic table. If they were included in the main body of the periodic table, the table would be much wider.

Using the periodic table, you can classify the elements in many ways, which the following sections describe. Two quite useful ways are

- ✔ **Metals, nonmetals, and metalloids:** These categories refer to large groupings of elements based on certain physical properties such as conductivity.

- ✔ **Families and periods:** These categories refer to the columns and rows, respectively, of the periodic table.

Classifying metals, nonmetals, and metalloids

Most of the elements in the periodic table are considered *metals.* The metals have properties that you normally associate with the metals you encounter in everyday life. They are solid (with the exception of mercury, Hg, a liquid), shiny, good conductors of electricity and heat, *ductile* (they can be drawn into thin wires), and *malleable* (they can be easily hammered into very thin sheets). And all these metals tend to lose electrons easily (see Chapter 13). Figure 5-2 shows the metals.

IA (1)	IIA (2)	IIIB (3)	IVB (4)	VB (5)	VIB (6)	VIIB (7)	VIIIB (8)	(9)	(10)	IB (11)	IIB (12)	13			
3 Li Lithium 6.939	4 Be Beryllium 9.0122														
11 Na Sodium 22.9898	12 Mg Magnesium 24.312											13 Al Aluminum 26.9815			
19 K Potassium 39.102	20 Ca Calcium 40.08	21 Sc Scandium 44.956	22 Ti Titanium 47.90	23 V Vanadium 50.942	24 Cr Chromium 51.996	25 Mn Manganese 54.9380	26 Fe Iron 55.847	27 Co Cobalt 58.9332	28 Ni Nickel 58.71	29 Cu Copper 63.546	30 Zn Zinc 65.37	31 Ga Gallium 69.72			
37 Rb Rubidium 85.47	38 Sr Strontium 87.62	39 Y Yttrium 88.905	40 Zr Zirconium 91.22	41 Nb Niobium 92.906	42 Mo Molybdenum 95.94	43 Tc Technetium (99)	44 Ru Ruthenium 101.07	45 Rh Rhodium 102.905	46 Pd Palladium 106.4	47 Ag Silver 107.868	48 Cd Cadmium 112.40	49 In Indium 114.82	50 Sn Tin 118.69		
55 Cs Cesium 132.905	56 Ba Barium 137.34	57 La Lanthanium 138.91	72 Hf Hafnium 179.49	73 Ta Tantalum 180.948	74 W Tungsten 183.85	75 Re Rhenium 186.2	76 Os Osmium 190.2	77 Ir Iridium 192.2	78 Pt Platinum 195.09	79 Au Gold 196.967	80 Hg Mercury 200.59	81 Tl Thallium 204.37	82 Pb Lead 207.19	83 Bi Bismuth 208.980	84 Po Polonium (210)
87 Fr Francium (223)	88 Ra Radium (226)	89 Ac Actinium (227)	104 Rf Rutherfordium (261)	105 Db Dubnium (262)	106 Sg Seaborgium (266)	107 Bh Bohrium (264)	108 Hs Hassium (269)	109 Mt Meitnerium (268)	110 Ds Darmstadtium (281)	111 Rg Roentgenium (286)	112 Cn Copernicium (285)				

58 Ce Cerium 140.12	59 Pr Praseodymium 140.907	60 Nd Neodymium 144.24	61 Pm Promethium (145)	62 Sm Samarium 150.35	63 Eu Europium 151.96	64 Gd Gadolinium 157.25	65 Tb Terbium 158.924	66 Dy Dysprosium 162.50	67 Ho Holmium 164.930	68 Er Erbium 167.26	69 Tm Thulium 168.934	70 Yb Ytterbium 173.04	71 Lu Lutetium 174.97
90 Th Thorium 232.038	91 Pa Protactinium (231)	92 U Uranium 238.03	93 Np Neptunium (237)	94 Pu Plutonium (242)	95 Am Americium (243)	96 Cm Curium (247)	97 Bk Berkelium (247)	98 Cf Californium (251)	99 Es Einsteinium (254)	100 Fm Fermium (257)	101 Md Mendelevium (258)	102 No Nobelium (259)	103 Lr Lawrencium (260)

Figure 5-2: The metals.

All the elements on the left-hand side and in the middle of the periodic table are metals, with one notable exception — hydrogen. It is the first element and is unique in its properties. Scientists stick it above lithium, but it doesn't react like a metal.

Meanwhile, except for a few elements that border the metals on the right (more on those in a second), the elements to the right of the periodic table are classified as *nonmetals* (along with hydrogen). These elements are shown in Figure 5-3.

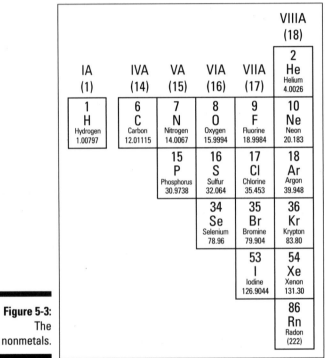

Figure 5-3:
The
nonmetals.

REMEMBER

Nonmetals have properties opposite those of the metals. The nonmetals are brittle, not malleable or ductile, poor conductors of both heat and electricity, and tend to gain electrons in chemical reactions. Some nonmetals are liquids; some are gases.

The elements that border the metals and the nonmetals are classified as *metalloids,* and they're shown in Figure 5-4.

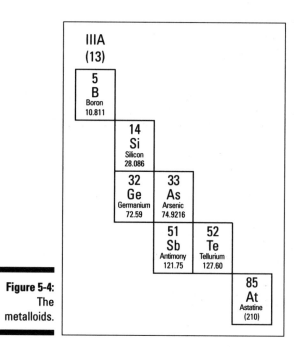

The metalloids, or *semimetals,* have properties that are somewhat of a cross between metals and nonmetals. They tend to be economically important because of their unique conductivity properties (they only partially conduct electricity), which make them valuable in the semiconductor and computer chip industry. (Did you think the term *silicon valley* referred to a valley covered in sand? Nope. Silicon, one of the metalloids, is used in making computer chips.)

Organizing by periods and families

The periodic table is composed of horizontal rows called *periods.* The periods are numbered 1 through 7 on the left-hand side of the table. The atomic numbers increase from left to right in each period.

Even though they're in the same period, these elements have chemical properties that are not all that similar. Consider the first two members of period 3: sodium (Na) and magnesium (Mg). In reactions, they both tend to lose electrons (after all, they are metals), but sodium loses one electron, whereas magnesium loses two. Chlorine (Cl), down near the end of the period, tends to gain an electron (it's a nonmetal).

The vertical columns are called *groups,* or *families.* The families may be labeled at the top of the columns in one of two ways. The older method uses Roman numerals and letters. Many chemists (especially old ones like me) prefer and still use this method. The newer method simply uses the numbers 1 through 18. I use the older method in describing the features of the table because it helps relate the position of an element on the periodic table with its number of valence electrons more than the 1–18 grouping system does.

The members of a family do have similar properties. Consider the IA family, starting with lithium (Li) — don't worry about hydrogen, because it's unique and doesn't really fit anywhere — and going through francium (Fr). All these elements tend to lose only one electron in reactions. And all the members of the VIIA family tend to gain one electron.

So why do the elements in the same family have similar properties? And why do some families have the particular properties of electron loss or gain? To find out, you can examine four specific families on the periodic table and look at the electron configurations for a few elements in each family.

My family name is special

Take a look at Figure 5-5, which lists some important families that are given special names:

✔ The IA family is made up of the *alkali metals.* In reactions, these elements all tend to lose a single electron. This family contains some important elements, such as sodium (Na) and potassium (K). Both of these elements play an important role in the chemistry of the body and are commonly found in salts.

✔ The IIA family is made up of the *alkaline earth metals.* All these elements tend to lose two electrons. Calcium (Ca) is an important member of the IIA family (you need calcium for healthy teeth and bones).

✔ The VIIA family is made up of the *halogens.* They all tend to gain a single electron in reactions. Important members in the family include chlorine (Cl), used in making table salt and bleach, and iodine (I). Tincture of iodine is sometimes used as a disinfectant.

✔ The VIIIA family is made up of the *noble gases.* These elements are *very* unreactive. For a long time, the noble gases were called the inert gases, because people thought that these elements wouldn't react at all. Later, a scientist named Neil Bartlett showed that at least some of the inert gases could be reacted, but they required very special conditions. After Bartlett's discovery, the gases were then referred to as noble gases.

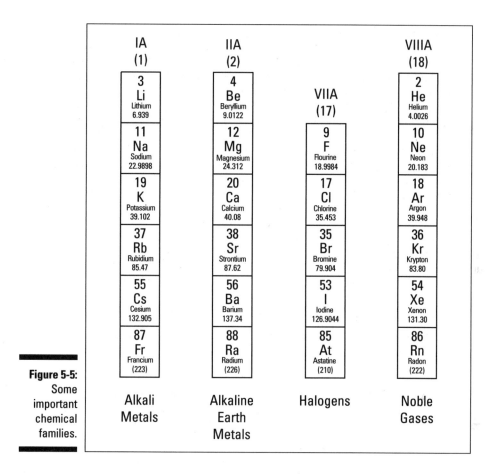

Figure 5-5:
Some
important
chemical
families.

What valence electrons have to do with families

Chapter 4 explains that an *electron configuration* shows the number of electrons in each orbital in a particular atom. The electron configuration forms the basis of the concept of bonding and molecular geometry and other important stuff that I cover in the various chapters of this book.

Tables 5-1 through 5-4 show the electron configurations for the first three members of the families IA, IIA, VIIA, and VIIIA.

Table 5-1	Electron Configurations for Members of IA (Alkali Metals)
Element	*Electron Configuration*
Li	$1s^22s^1$
Na	$1s^22s^22p^63s^1$
K	$1s^22s^22p^63s^23p^64s^1$

Table 5-2	Electron Configurations for Members of IIA (Alkaline Earth Metals)
Element	*Electron Configuration*
Be	$1s^22s^2$
Mg	$1s^22s^22p^63s^2$
Ca	$1s^22s^22p^63s^23p^64s^2$

Table 5-3	Electron Configurations for Members of VIIA (Halogens)
Element	*Electron Configuration*
F	$1s^22s^22p^5$
Cl	$1s^22s^22p^63s^23p^5$
Br	$1s^22s^22p^63s^23p^64s^23d^{10}4p^5$

Table 5-4	Electron Configurations for Members of VIIIA (Noble Gases)
Element	*Electron Configuration*
Ne	$1s^22s^22p^6$
Ar	$1s^22s^22p^63s^23p^6$
Kr	$1s^22s^22p^63s^23p^64s^23d^{10}4p^6$

These electron configurations show that each group of elements has some similarities in terms of the elements' valence electrons. *Valence electrons* are the s and p electrons in the outermost energy level of an atom (see Chapter 4).

Look at the electron configurations for the alkali metals (Table 5-1). In lithium, energy level 1 is filled, and a single electron is in the 2s orbital. In sodium, energy levels 1 and 2 are filled, and a single electron is in energy level 3. All these elements have one valence electron in an s orbital. The alkaline earth elements (Table 5-2) each have two valence electrons. The halogens (Table 5-3) each have seven valence electrons (in s and p orbitals — d orbitals don't count), and the noble gases (Table 5-4) each have eight valence electrons, which fill their valence orbitals.

So how do you remember all this stuff? Well, for the families labeled with a Roman numeral and an A, the Roman numeral gives the number of valence electrons. The *IA* family has *1* valence electron, the *IIA* family has *2* valence electrons, the *VIIA* family has *7* valence electrons, and the *VIIIA* family has *8* valence electrons. Pretty cool, eh?

The Roman numeral makes it very easy to determine that oxygen (O) has six valence electrons (it's in the VIA family), that silicon (Si) has four, and so on. You don't even have to write the electronic configuration or the energy diagram to determine the number of valence electrons.

Noble and gassy

The fact that the noble gases have eight valence electrons, filling their valence (outermost) energy level, explains why the noble gases are extremely hard to react. A filled (complete) valence level makes an element stable. It doesn't easily lose, gain, or share electrons.

A lot of stability in nature seems to be associated with this condition. Chemists observe that the other elements in the A families on the periodic table tend to lose, gain, or share valence electrons in order to achieve the goal of having a filled valence shell of eight electrons.

Chemists sometimes called this phenomenon the *octet rule.* For example, look at the electron configuration for sodium (Na): $1s^2 2s^2 2p^6 3s^1$. It has one valence electron — the $3s^1$. If it lost that electron, its valence shell would be energy level 2, which is filled. Without the $3s^1$, it would become *isoelectronic* with neon (Ne) (meaning the two elements have the same electronic configuration) and achieve stability. As I show you in Chapters 13 and 14, achieving stability by having a filled valence shell is the driving force in chemical bonding.

But what about elements that are labeled with a Roman numeral and a B? These elements, found in the middle of the periodic table, are commonly called the *transition metals*. Their electrons progressively fill the d orbitals. Scandium (Sc) is the first member of the transition metals, and it has an electronic configuration of $1s^22s^22p^63s^23p^64s^23d^1$. Titanium (Ti), the next transition metal, has a configuration $1s^22s^22p^63s^23p^64s^23d^2$. Notice that the number of electrons in the s and p orbitals doesn't change, but the progressively added electrons fill the d orbitals. Lanthanides and actinides, the two groups of elements that are pulled out of the main body of the periodic table and shown below it, are classified as *inner transition metals*. In these elements, the electrons progressively fill the f orbitals in much the same way that the electrons of the transition metals fill the d orbitals.

Chapter 6

Balloons, Tires, and Scuba Tanks: The Wonderful World of Gases

*G*ases are all around you. Because gases are generally invisible, you may not think of them directly, but you're certainly aware of their properties. You breathe a mixture of gases that you call air. You check the pressure of your automobile tires, and you check the atmospheric pressure to see if a storm is coming. You burn gases in your gas grill and lighters. You fill birthday balloons with gases for your loved ones.

The properties of gases and their interrelationships are important to you. Is there enough pressure in my tires? How big is that balloon going to be? Do my scuba tanks have enough air? The list goes on and on.

In this chapter, I introduce you to gases at both the microscopic and macroscopic levels. I show you one of science's most successful theories — the kinetic molecular theory of gases. I explain the macroscopic properties of gases and show you the important interrelationships among them. I also show you how these relationships come into play in reaction stoichiometry. This chapter is a real gas!

Taking a Microscopic View of Gases: The Kinetic Molecular Theory

A theory is useful to scientists if it describes the physical system they're examining and allows them to predict what will happen if they change some variable. The *kinetic molecular theory of gases* does just that. It has limitations — all theories do — but it's one of the most useful theories in chemistry. This section describes the theory's basic *postulates* — assumptions, hypotheses, axioms (pick your favorite word) you can accept as being true.

✔ **Gases are composed of tiny particles, either atoms or molecules.**

Unless you're discussing matter at greatly elevated temperatures, the particles referred to as gases tend to be relatively small. The more massive particles clump together to form liquids or even solids. So gases are normally small with relatively low atomic and molecular weights.

✔ **The gas particles are so small when compared to the distances between them that the volume the gas particles themselves take up is negligible and is assumed to be zero.**

These gas particles do take up some volume — that's one of the properties of matter. But the gas particles are small, so if a container doesn't hold many of them, you say that their volume is negligible when compared to the volume of the container or the space between the gas particles. Because of all that space between the gas particles, they can be squeezed together to compress the gas. Solids and liquids can't be squeezed, because their particles are *much* closer together. (Chapter 3 covers the various states of matter, if you want to have a look-see at the differences between solids, liquids, and gases.)

The concept of a *negligible* quantity is used a lot in chemistry. An example is in Chapter 11 where I use the acid ionization constant (K_a) of a weak acid, ignoring the amount of weak acid that has ionized compared to the initial concentration of the acid.

In the real world, I like to compare this negligible concept to finding a dollar in the street. If you have no money at all, then that dollar represents a sizable quantity of cash (perhaps your next meal). But if you're a multimillionaire, then that dollar doesn't represent much at all. It may as well be a piece of scrap paper. You may not even pick it up. (I really can't imagine being that rich.) Its value is negligible when compared to the rest of your wealth. And sure, the gas particles have a volume, but it's so small that it's insignificant when compared with the distance between the gas particles and the volume of the container.

✔ **The gas particles are in constant random motion, moving in straight lines and colliding with the inside walls of the container.**

The gas particles are always moving in a straight-line motion. (Gases have a higher kinetic energy — energy of motion — associated with them than solids or liquids do; see Chapter 3.) They continue to move in these straight lines until they collide with something — either with each other or with the inside walls of the container. The particles also all move in different directions, so the collisions with the inside walls of the container tend to be uniform over the entire inside surface.

You can observe this uniformity by simply blowing up a balloon. The balloon is spherical because the gas particles are hitting all points of the inside walls the same. The collision of the gas particles with the inside walls of the container is called *pressure.* The idea that the gas particles are in constant, random, straight-line motion explains why gases uniformly mix if put in the same container. It also explains why, when you drop a bottle of cheap perfume at one end of the room, the people at the other end of the room are able to smell it right away.

✔ **The gas particles are assumed to have negligible attractive or repulsive forces between each other.**

In other words, the gas particles are assumed to be totally independent, neither attracting nor repelling each other. That said, it's hair-splitting time: This assumption is actually false. If it were true, chemists would never be able to liquefy a gas, which they can. But the reason you can accept this assumption as true (or at least useful) is that the attractive and repulsive forces are generally so small that they can safely be ignored. The assumption is most valid for nonpolar gases, such as hydrogen and nitrogen, because the attractive forces involved are London forces. However, if the gas molecules are polar, as in water and HCl, this assumption can become a problem. (Turn to Chapter 17 for the scoop about London forces and polar things — all related to the attraction between molecules.)

✔ **The gas particles may collide with each other. These collisions are assumed to be elastic, with the total amount of kinetic energy of the two gas particles remaining the same.**

Not only do the gas particles collide with the inside walls of the container, but they also collide with each other. If they hit each other, no kinetic energy is lost, but kinetic energy may be transferred from one gas particle to the other. For example, imagine two gas particles — one moving fast and the other moving slow — colliding. Kinetic energy is transferred from the faster particle to the slower particle. The slow-moving particle bounces off the faster particle and moves away at a greater speed than before, while the faster particle bounces off the slower particle and moves away at a slower speed. The total amount of kinetic energy remains the same, but one gas particle loses energy and the other gains energy. This transfer of energy is the principle behind pool — you transfer kinetic energy from your moving pool stick to the cue ball to the ball you're aiming at.

✔ **The Kelvin temperature is directly proportional to the *average* kinetic energy of the gas particles.**

The gas particles aren't all moving with the same amount of kinetic energy. A few are moving relatively slow and a few are moving very fast, but most are somewhere in between these two extremes. Temperature, particularly as measured using the Kelvin temperature scale, is directly related to the *average* kinetic energy of the gas. If you heat the gas so that the Kelvin temperature (K) increases, the average kinetic energy of the gas also increases. (To calculate the Kelvin temperature, add 273 to the Celsius temperature: K = °C + 273. Temperature scales and average kinetic energy are all tucked neatly into Chapter 3.)

A gas that obeys all the postulates of the kinetic molecular theory is called an *ideal gas*. Obviously, no real gas obeys the assumptions made in the second and fourth postulates *exactly* (all gas particles actually do have small measures of volume and attractive or repulsive force). But a nonpolar gas at high temperatures and low pressure (concentration) approaches ideal gas behavior.

Staying Under Pressure — Atmospheric Pressure, That Is

Although you're not in a container, the gas molecules of the atmosphere are constantly hitting you, your books, your computer, and everything and exerting a force called *atmospheric pressure*. Atmospheric pressure is measured using an instrument called a *barometer*. The following sections give you the lowdown on atmospheric pressure.

Measuring atmospheric pressure: The barometer

If you get a complete weather report, the atmospheric pressure is normally included. You can get an idea about changes in the weather by observing whether the atmospheric pressure is rising or falling. The atmospheric pressure is measured using a barometer, and Figure 6-1 shows the components of one.

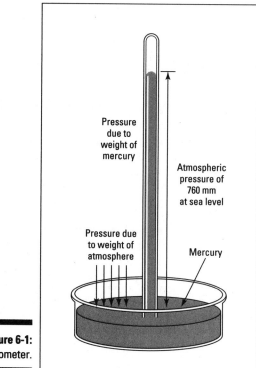

Pressure
due to
weight of
mercury

Atmospheric
pressure of
760 mm
at sea level

Pressure due
to weight of
atmosphere

Mercury

Figure 6-1:
A barometer.

A barometer is composed of a long glass tube that's closed at one end and totally filled with a liquid. You can use water, but the tube would have to be *very* long (about 35 feet long), making for a rather inconvenient barometer. Barometers typically use mercury because it's a very dense liquid. The tube filled with mercury is inverted into an open container of mercury so that the open end of the tube is under the surface of the mercury in the container. The force of gravity pulls the mercury in the tube *down*, causing it to drain out into the container, while the weight of the gases in the atmosphere exerts pressure downward on the mercury in the open container and forces it *up* into the tube. Sooner or later, these forces balance, and the mercury in the tube comes to rest at a certain height in the tube. The greater the pressure of the atmosphere, the higher the mercury column that can be measured; the lower the pressure of the atmosphere (for example, at the top of a tall mountain), the shorter the column. At sea level, the column is 760 millimeters high, the so-called *normal* atmospheric pressure.

Atmospheric pressure can be expressed a number of different ways. It can be expressed in millimeters of mercury (mm Hg); atmospheres (atm), a unit of pressure where 1 atmosphere is the pressure at sea level; torr, a unit of pressure where 1 torr equals 1 millimeter of mercury; pounds per square inch (psi); pascals (Pa), a unit of pressure where 1 pascal equals 1 newton per square meter (don't worry about what a newton is; it's just a way to express pressure); or kilopascals (kPa), where 1 kilopascal equals 1,000 pascals.

So you can express the atmospheric pressure at sea level as

$$760 \text{ mm Hg} = 1 \text{ atm} = 760 \text{ torr} = 14.69 \text{ psi} = 101,325 \text{ Pa} = 101.325 \text{ kPa}$$

Note that sometimes you also hear atmospheric pressure reported as inches of mercury (1 atm = 29.921 in Hg). In this book, I primarily use atmospheres and torr, with an occasional millimeter of Hg. Variety is the spice of life.

Measuring confined gas pressure: The manometer

You can measure the pressure of a gas confined in a container by using an apparatus called a *manometer* (pronounced man-*ah*-muh-ter). Figure 6-2 shows the components of a manometer.

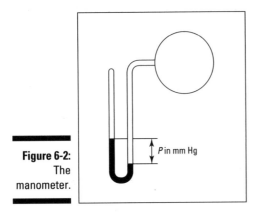

Figure 6-2:
The
manometer.

P in mm Hg

A manometer is kind of like a barometer. The container of gas is attached to a U-shaped piece of glass tubing that's partially filled with mercury and sealed at the other end. Gravity pulls down the mercury column at the closed end. The mercury is then balanced by the pressure of the gas in the container. The difference in the two mercury levels represents the amount of gas pressure.

Grasping Different Gas Laws

Various scientific laws describe the relationships among four of the important physical properties of gases:

- Volume
- Pressure
- Temperature
- Amount

This section covers those various laws. Boyle's, Charles's, and Gay-Lussac's laws each describe the relationship between two properties while keeping the other two properties constant. (In other words, you take two properties, change one, and then see its effect on the second — while keeping the remaining properties constant.) Another law — a combo of Boyle's, Charles's, and Gay-Lussac's individual laws — enables you to vary more than one property at a time.

But that combo law doesn't let you vary the physical property of amount. Avogadro's law, however, does. And an ideal gas law even lets you take into account variations in all four physical properties.

Yes, this section is so chock-full of laws, it'll probably *give* you gas just trying to digest it.

Boyle's law: Nothing to do with boiling

Boyle's law, named after Robert Boyle, a 17th-century English scientist, describes the pressure-volume relationship of gases if the temperature and amount are kept constant. Figure 6-3 illustrates the pressure-volume relationship using the kinetic molecular theory.

The left-hand cylinder in the figure contains a certain volume of gas at a certain pressure. When the volume is decreased, the same number of gas particles is now contained in a much smaller volume and the number of collisions increases significantly. Therefore, the pressure is greater.

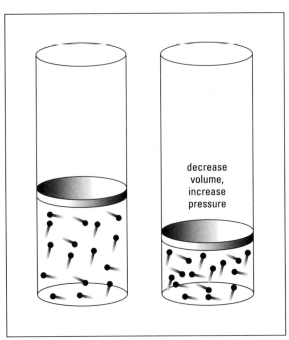

Figure 6-3:
Pressure-
volume
relationship
of gases —
Boyle's law.

Boyle's law states that an inverse relationship exists between the volume and the pressure. As the volume decreases, the pressure increases, and vice versa. Boyle determined that the product of the pressure and the volume is a constant (k):

$$PV = k$$

Now consider a case where you have a gas at a certain pressure (P_1) and volume (V_1). If you change the volume to some new value (V_2), the pressure also changes to a new value (P_2). You can use Boyle's law to describe both sets of conditions:

$$P_1 V_1 = k$$
$$P_2 V_2 = k$$

The constant, k, is the same in both cases. So you can say

$$P_1 V_1 = P_2 V_2 \qquad \text{(with temperature and amount constant)}$$

This equation is another statement of Boyle's law — and it's really a more useful one, because you'll normally deal with changes in pressure and volume. If you know three of the preceding quantities, you can calculate the fourth one. For example:

Suppose that you have 5.00 liters of a gas at 1.00 atm pressure, and then you decrease the volume to 2.00 liters. What's the new pressure?

To find the answer, follow these steps:

1. **Use the following setup:**

$$P_1V_1 = P_2V_2$$

2. **Substitute 1.00 atmospheres for P_1, 5.00 liters for V_1, and 2.00 liters for V_2 and you get**

$$(1.00 \text{ atm})(5.00 \text{ L}) = P_2(2.00 \text{ L})$$

3. **Solve for P_2:**

$$\frac{(1.00 \text{ atm})(5.00 \text{ L})}{2.00 \text{ L}} = P_2 = 2.50 \text{ atm}$$

The answer makes sense, because you decreased the volume and the pressure increased, which is exactly what Boyle's law says.

Charles's law: Don't call me Chuck

Charles's law, named after Jacques Charles, a 19th-century French chemist, has to do with the relationship between volume and temperature, keeping the pressure and amount constant. You run across situations dealing with this relationship in everyday life, especially in terms of the heating and cooling of balloons.

Figure 6-4 shows the temperature-volume relationship.

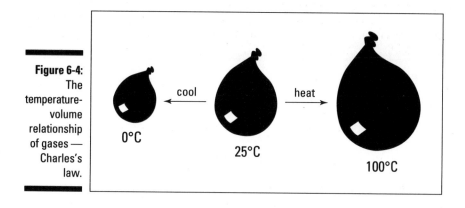

Figure 6-4: The temperature-volume relationship of gases — Charles's law.

cool

heat

0°C

25°C

100°C

Look at the balloon in the middle of Figure 6-4. What do you think would happen to the balloon if you placed it in the freezer or took it outside in subzero weather? It'd get smaller. Inside the freezer or in arctic weather, the external pressure, or atmospheric pressure, is the same, but the gas particles inside the balloon aren't moving as fast, so the volume shrinks to keep the pressure constant. If you heat the balloon, the balloon expands and the volume increases. This correspondence is a *direct relationship* — as the temperature increases, the volume increases, and vice versa.

Jacques Charles developed the mathematical relationship between temperature and volume. He also discovered that you must use the Kelvin (K) temperature when working with gas law expressions and calculations.

Charles's law says that the volume is directly proportional to the Kelvin temperature. Mathematically, the law looks like this:

$$V = bT \text{ or } \frac{V}{T} = b$$

(where b is a constant)

If the temperature of a gas with a certain volume (V_1) and Kelvin temperature (T_1) is changed to a new Kelvin temperature (T_2), the volume also changes (V_2).

$$\frac{V_1}{T_1} = b \quad \frac{V_2}{T_2} = b$$

The constant, b, is the same, so

$$\frac{V_1}{T_1} = \frac{V_2}{T_2}$$

(with the pressure and amount of gas held constant and temperature expressed in K).

If you have any three of the quantities, you can calculate the fourth. For example:

Suppose you live in Alaska and are outside in the middle of winter, where the temperature is –23 degrees Celsius. You blow up a balloon so that it has a volume of 1.00 liter. You then take it inside your home, where the temperature is a toasty 27 degrees Celsius. What's the new volume of the balloon?

To solve, follow these steps:

1. **Convert your temperatures to Kelvin by adding 273 to the Celsius temperature:**

   $$-23°C + 273 = 250K \qquad \text{(outside)}$$
   $$27°C + 273 = 300K \qquad \text{(inside)}$$

2. **Solve for V_2, using the following setup:**

   $$\frac{V_1}{T_1} = \frac{V_2}{T_2}$$

3. **Multiply both sides by T_2 so that V_2 is on one side of the equation by itself:**

   $$\frac{V_1 T_2}{T_1} = V_2$$

4. **Substitute the values to calculate the following answer:**

   $$\frac{(1.00 \text{ L})(300.K)}{250K} = V_2 = 1.20 \text{ L}$$

This answer is reasonable, because Charles's law says that if you increase the Kelvin temperature, the volume increases.

Gay-Lussac's law

Gay-Lussac's law, named after the 19th-century French scientist Joseph-Louis Gay-Lussac, deals with the relationship between the pressure and temperature of a gas if its volume and amount are held constant. Imagine, for example, that you have a metal tank of gas. The tank has a certain volume, and the gas inside has a certain pressure. If you heat the tank, you increase the kinetic energy of the gas particles. So they're now moving much faster, and they're not only hitting the inside walls of the tank more often but also hitting with more force. The pressure has increased.

Gay-Lussac's law says that the pressure is directly proportional to the Kelvin temperature. Figure 6-5 shows this relationship.

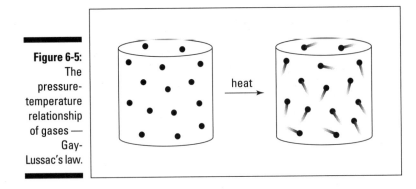

Figure 6-5:
The pressure-temperature relationship of gases — Gay-Lussac's law.

heat

Mathematically, Gay-Lussac's law is represented by the following equation, where P is the pressure, T is the temperature, and k is a constant:

$$P = kT \left(\text{or } \frac{P}{T} = k \right)$$

(at constant volume and amount)

Consider a gas at a certain Kelvin temperature and pressure (T_1 and P_1), with the conditions being changed to a new temperature and pressure (T_2 and P_2):

$$\frac{P_1}{T_1} = \frac{P_2}{T_2}$$

EXAMPLE

If you have a tank of gas at 800 torr pressure and a temperature of 250 Kelvin, and it's heated to 400 Kelvin, what's the new pressure?

To find the new pressure, follow these steps:

1. **Starting with** $\frac{P_1}{T_1} = \frac{P_2}{T_2}$, **multiply both sides by** T_2 **so you can solve for** P_2:

$$\frac{P_1 T_2}{T_1} = P_2$$

2. **Substitute the values to calculate the following answer:**

$$\frac{(800. \text{ torr})(400.\text{K})}{250\text{K}} = P_2 = 1,280 \text{ torr}$$

This answer is reasonable because if you heat the tank, the pressure should increase.

The combined gas law

All the examples in the preceding sections assume that two properties are held constant and one property is changed to see its effect on a fourth property. But life is rarely that simple. How do you handle situations in which two or even three properties change? You can treat each one separately, but it sure would be nice if you had a way to combine things so that wouldn't be necessary.

Actually, there is a way. You can combine Boyle's law, Charles's law, and Gay-Lussac's law into one equation. Trust me, you don't want me to show you exactly how it's done because it involves a lot of boring algebra, but the end result is called the *combined gas law,* and it looks like this:

$$\frac{P_1V_1}{T_1} = \frac{P_2V_2}{T_2}$$

Just like in the preceding examples, *P* is the pressure of the gas (in atm, mm Hg, torr, and so on), *V* is the volume of the gas (in appropriate units), and *T* is the temperature (in Kelvin). The *1* and *2* stand for the initial and final conditions, respectively. The amount is still held constant: No gas is added, and no gas escapes. Six quantities are involved in this combined gas law, so knowing five allows you to calculate the sixth.

Suppose that a weather balloon with a volume of 25.0 liters at 1.00 atmospheres pressure and a temperature of 27 degrees Celsius is allowed to rise to an altitude where the pressure is 0.500 atmospheres and the temperature is –33 degrees Celsius. What's the new volume of the balloon?

Before I show you how to work this problem, do a little reasoning. The temperature is decreasing, so that change should cause the volume to decrease (Charles's law). However, the pressure is also decreasing, which should cause the balloon to expand (Boyle's law). These two factors are competing, so at this point, you don't know which will win out.

You're looking for the new volume (V_2), so solve as such:

1. **Rearrange the combined gas law to obtain the following equation (by multiplying each side by T_2 and dividing each side by P_2, which puts V_2 by itself on one side):**

$$\frac{P_1V_1T_2}{P_2T_1} = V_2$$

2. Identify your quantities:

$$P_1 = 1.00 \text{ atm}, V_1 = 25.0 \text{ L}, T_1 = 27°C + 273 = 300.\text{K}$$
$$P_2 = 0.500 \text{ atm}, T_2 = -33°C + 273 = 240.\text{K}$$

3. Substitute the values to calculate the following answer:

$$\frac{(1.00 \text{ atm})(25.0 \text{ L})(240.\text{K})}{(0.500 \text{ atm})(300.\text{K})} = V_2$$

$$40.0 \text{ L} = V_2$$

Because the volume increased overall in this case, Boyle's law had a greater effect than Charles's law.

Avogadro's law

The combined gas equation gives you a way to calculate changes involving pressure, volume, and temperature. But you still have the problem of amount to deal with. In order to account for amount, you need to know another law.

Amedeo Avogadro (the same Avogadro that gave us his famous number of particles per mole — see Chapter 8) determined, from his study of gases, that equal volumes of gases at the same temperature and pressure contain equal numbers of gas particles. So Avogadro's law says that the volume of a gas is directly proportional to the number of moles of gas (number of gas particles) at a constant temperature and pressure. Mathematically, Avogadro's law looks like this:

$$V = kn \text{ (at constant temperature and pressure)}$$

In this equation, k is a constant and n is the number of moles of gas. If you have a number of moles of gas (n_1) at one volume (V_1), and the moles change due to a reaction (n_2), the volume also changes (V_2), giving you the equation

$$\frac{V_1}{n_1} = \frac{V_2}{n_2}$$

I don't work any problems with this law here, because it's basically the same idea as the other gas laws covered in this chapter.

A very useful consequence of Avogadro's law is that the volume of a mole of gas can be calculated at any temperature and pressure. An extremely useful form to know when calculating the volume of a mole of gas is *1 mole of any gas at STP occupies 22.4 liters. STP* in this case is not an oil or gas additive. It stands for *standard temperature and pressure.*

- ✔ **Standard pressure:** 1.00 atm (760. torr or 760. mm Hg)
- ✔ **Standard temperature:** 273K

This relationship between moles of gas and liters gives you a way to convert the gas from a mass to a volume.

Suppose that you have 50.0 grams of oxygen gas (O_2), and you want to know its volume at STP.

You can set up the problem like this (see Chapter 8 for the nuts and bolts of using moles in chemical equations):

$$\frac{50.0 \text{ g } O_2}{1} \times \frac{1 \text{ mol } O_2}{32.0 \text{ g } O_2} \times \frac{22.4 \text{ L}}{1 \text{ mol } O_2} = 35.0 \text{ L}$$

You now know that the 50.0 grams of oxygen gas occupies a volume of 35.0 liters at STP. But what if the gas is not at STP?

What's the volume of 50.0 grams of oxygen at 2.00 atm and 27.0 degrees Celsius?

In the next section, I show you a really easy way of doing this problem. But right now, you can use the combined gas law, because you know the volume at STP:

$$\frac{P_1 V_1}{T_1} = \frac{P_2 V_2}{T_2}$$

P_1 = 1.00 atm, V_1 = 35.0 L, T_1 = 273K

P_2 = 2.00 atm, T_2 = 300.K (27°C + 273)

Solving for V_2, you calculate the following answer:

$$\frac{P_1 V_1 T_2}{P_2 T_1} = V_2$$

$$\frac{(1.00 \text{ atm})(35.0 \text{ L})(300.\text{K})}{(2.00 \text{ atm})(273\text{K})} = V_2 = 19.2 \text{ L}$$

The ideal gas equation

If you take Boyle's law, Charles's law, Gay-Lussac's law, and Avogadro's law and throw them into a blender on high for a minute, you get the *ideal gas equation* — a way of working in volume, temperature, pressure, *and* amount. The ideal gas equation has the following form:

$$PV = nRT$$

The *P* represents pressure in atmospheres (atm), the *V* represents volume in liters (L), the *n* represents moles of gas, the *T* represents the temperature in Kelvin (K), and the R represents the ideal gas constant, which is 0.0821 liters atm/K mol.

Using this value of the ideal gas constant, the pressure must be expressed in atm, and the volume must be expressed in liters. You can calculate other ideal gas constants if you really want to use torr and milliliters, for example, but why bother? It's easier to memorize one value for R and then remember to express the pressure and volume in the appropriate units. Naturally, you'll *always* express the temperature in Kelvin when working any kind of gas law problem.

That said, now I want to show you an easy way to convert a gas from a mass to a volume if the gas is not at STP:

> What's the volume of 50.0 grams of oxygen at 2.00 atm and 27.0 degrees Celsius?

To answer this question, do the following:

1. **Convert the 50.0 grams of oxygen to moles using the molecular weight of O_2:**

$$\frac{50.0 \text{ g}}{1} \times \frac{1 \text{ mol}}{32.0 \text{ g}} = 1.562 \text{ mol}$$

2. **Take the ideal gas equation and rearrange it so you can solve for *V*:**

$$PV = nRT$$

$$V = \frac{nRT}{P}$$

3. **Add your known quantities to calculate the following answer:**

$$\frac{(1.562 \text{ mol})\left(0.0821 \frac{\text{L atm}}{\text{K mol}}\right)(300.\text{K})}{2.00 \text{ atm}} = 19.2 \text{ L}$$

This answer is exactly the same as what you get in the preceding section, but it's calculated in a much more straightforward way.

The van der Waals equation

The ideal gas equation is useful for gases that come close to obeying the kinetic molecular theory; in other words, an ideal gas obeys the ideal gas equation. But what about those situations in which the gas isn't ideal and doesn't obey the ideal gas equation? This is true of polar gases (which have increased attractive forces) or high concentrations of gases (for which you can't really ignore the volume of the gas particles). Johannes van der Waals developed an equation that takes into account the attractive forces and the volumes of real gases by introducing two constants, a and b, into the ideal gas equation. The values for the a and b constants for each individual gas may be found in the Chemical Rubber Company (CRC) Handbook. The resulting equation is called the *van der Waals equation:*

$$\left(P + \frac{an^2}{V^2}\right)(V - nb) = nRT$$

The attraction of the gas particles for each other slightly reduces the pressure of the gas. The a constant, combined with the moles of gas, *n,* adjusts for the attractive force:

$$\left(P + \frac{an^2}{V^2}\right)$$

The greater the attractive force, the greater the value of a.

The actual volume of a real gas is less than an ideal gas. The volume can be corrected by the $V - nb$ term. The larger the gas particle, the more volume it occupies and the large its b value.

Applying Gas Laws to Stoichiometry

The ideal gas equation (and even the combined gas equation) allows chemists to work stoichiometry problems involving gases. (Chapter 8 is your key to the world of stoichiometry.) In this section, you use the ideal gas equation to do such a problem, using a classic chemistry experiment — the decomposition of potassium chlorate to potassium chloride and oxygen by heating:

$$2\ KClO_3(s) \rightarrow 2\ KCl(s) + 3\ O_2(g)$$

Figure out the volume of oxygen gas produced at 700 torr and 27 degrees Celsius from the decomposition of 25.0 grams of $KClO_3$.

To calculate the volume, do the following:

1. **Calculate the number of moles of oxygen gas produced:**

$$\frac{25.0 \text{ g } KClO_3}{1} \times \frac{1 \text{ mol } KClO_3}{122.55 \text{ g } KClO_3} \times \frac{3 \text{ mol } O_2}{2 \text{ mol } KClO_3} = 0.3060 \text{ mol } O_2$$

2. **Convert the temperature to Kelvin and the pressure to atmospheres:**

$27°C + 273 = 300.K$

$700 \text{ torr}/760 \text{ torr/atm} = 0.9211 \text{ atm}$

3. **Put everything in the ideal gas equation:**

$PV = nRT$

$V = \dfrac{nRT}{P}$

$$\frac{(0.3060 \text{ mol})\left(0.0821 \frac{\text{L atm}}{\text{K mol}}\right)(300.K)}{0.9211 \text{ atm}} = 8.18 \text{ L}$$

Tackling Dalton's and Graham's Laws

This section covers a couple of miscellaneous but fine gas laws you should have a nodding acquaintance with. One relates to partial pressures and the other to gaseous effusion/diffusion. Party on.

Dalton's law

Dalton's law of partial pressures says that in a mixture of gases, the total pressure is the sum of the partial pressures of each individual gas.

If you have a mixture of gases — gas A, gas B, gas C, and so on — then the total pressure of the system is simply the sum of the pressures of the individual gases. Mathematically, the relationship can be expressed like this:

$P_{\text{total}} = P_A + P_B + P_C + \dots$

When working stoichiometry problems like the one in the preceding section involving the decomposition of potassium chlorate, the oxygen is normally collected over water by displacement and the volume is then measured.

However, in order to get the pressure of just the oxygen, you have to subtract the pressure due to the water vapor. You have to mathematically "dry out" the gas.

Suppose, for example, that a sample of oxygen is collected over water at a total pressure of 755 torr at 20 degrees Celsius. And suppose that your job, you lucky dawg, is to calculate the pressure of the oxygen.

You know that the total pressure is 755 torr. Your first task is to reference a table of vapor pressures of water versus temperature. (You can find such a table in a variety of places, such as the Chemical Rubber Company (CRC) Handbook.) After looking at the table, you determine that the partial pressure of water at 20 degrees Celsius is 17.5 torr. Now you're ready to calculate the pressure of the oxygen:

$$P_{total} = P_{oxygen} + P_{water\ vapor}$$

$$755\ torr = P_{oxygen} + 17.5\ torr$$

$$P_{oxygen} = 755\ torr - 17.5\ torr = 737.5\ torr$$

Knowing the partial pressure of gases like oxygen is important in deep-sea diving and the operation of respirators in hospitals.

Graham's law

Place a few drops of a strong perfume on a table at one end of a room, and soon people at the other end of the room can smell it. This process is called *gaseous diffusion,* the mixing of gases due to their molecular motion.

Place a few drops of that same perfume inside an ordinary rubber balloon and blow it up. Very soon you'll be able to smell the perfume outside of the balloon as it makes its way through the microscopic pores of the rubber. This process is called *gaseous effusion,* the movement of a gas through a tiny opening. The same process of effusion is responsible for the helium being quickly lost from rubber balloons.

Thomas Graham determined that the rates or velocities (v) of diffusion and effusion of gases are inversely proportional to the square roots of their molecular or atomic weights (M). This relationship is called Graham's law. In general, it says that the lighter the gas, the faster it will effuse (or diffuse). Mathematically, Graham's law looks like this:

$$\frac{v_1}{v_2} = \sqrt{\frac{M_2}{M_1}}$$

The sweet smell of diffusion

The concept of diffusion can be easily demonstrated by using latex balloons and some flavoring extract. Place a few drops of an extract (vanilla, peppermint, almond, or whatever you like) inside a balloon. Then inflate the balloon, tie it off, and shake it. After a few minutes, smell the balloon. You should be able to detect the specific aroma of the extract you placed inside the balloon.

The extracts are fairly volatile substances, and they vaporize inside the balloon and diffuse through the pores in the latex to our noses. Some chemists believe that the extract molecules may actually interact with the latex, making penetration of the balloon walls easier.

Suppose that you fill two rubber balloons to the same size, one with hydrogen (H_2) and the other with oxygen (O_2). The hydrogen, being lighter, should effuse through the balloon pores faster. But how much faster? Using Graham's law, you can determine the answer:

$$\frac{v_{H_2}}{v_{O_2}} = \sqrt{\frac{M_{O_2}}{M_{H_2}}}$$

$$\frac{v_{H_2}}{v_{O_2}} = \sqrt{\frac{32.0 \text{ g/mol}}{2.0 \text{ g/mol}}}$$

$$\frac{v_{H_2}}{v_{O_2}} = \sqrt{16}$$

$$\frac{v_{H_2}}{v_{O_2}} = 4$$

The hydrogen should effuse out four times as fast as the oxygen.

Part II
A Cornucopia of Chemical Concepts

In this part . . .

Mention chemistry, and most people immediately think of chemical reactions. Scientists use chemical reactions to make new drugs, plastics, cleaners, fabrics — the list is endless. They also use chemical reactions to analyze samples and find out what and how much is in them. Chemical reactions power our bodies, our sun, and our universe. Chemistry is all reactions and the bonding that occurs in them. And those reactions and bonds are what this part is all about.

These chapters introduce you to chemical reactions. I then introduce you to the mole. No, not the small furry animal that burrows holes in your backyard, but the concept that unites the microscopic world of atoms and molecules with the macroscopic world of grams and metric tons. In this part I also show you solutions, how to make them and how to calculate their concentrations. To heat things up, I introduce you to thermochemistry. Chemical reactions either give off energy or absorb energy, and in the thermochemistry chapter I show you how to calculate how much heat is given off or absorbed. Finally, I cover acids and bases, discussing their properties, including their sour and bitter tastes, along with the concept of pH. I don't think this part will leave a sour taste in your mouth. In fact, I don't see how you can fail to react to it.

Chapter 7

Chemical Cooking: Chemical Reactions

hemists do a lot of things: They measure the physical properties of substances, they analyze mixtures to find out what they're composed of, and they make new substances. The process of making chemical compounds is called *synthesis,* and it depends on chemical reactions. I always thought that being a synthetic organic chemist and working on the creation of new and potentially important compounds would be neat. I can just imagine the thrill of working for months, or even years, and finally ending up with a little pile of "stuff" that nobody in the world has ever seen. Hey, I am a nerd, after all!

In this chapter, I discuss chemical reactions — how they occur, the different types of chemical reactions, and how to write a balanced chemical equation.

Knowing What You Have and What You'll Get: Reactants and Products

In a chemical reaction, substances (elements and/or compounds) are changed into other substances (compounds and/or elements). You can't change one element into another in a chemical reaction — that happens in nuclear reactions, as I describe in Chapter 20. Instead, you create new substances with chemical reactions.

A number of clues show that a chemical reaction has taken place — something new is visibly produced, a gas is created, heat is given off or taken in, and so on. The chemical substances that are eventually changed are called the *reactants,* and the new substances that are formed are called the *products. Chemical equations* show the reactants and products, as well as other factors such as energy changes, catalysts, and so on. With these equations, an arrow is used to indicate that a chemical reaction has taken place. In general terms, a chemical reaction follows this format:

Reactants → Products

For example, take a look at the reaction that occurs when you light your natural gas range in order to fry your breakfast eggs. Methane (natural gas) reacts with the oxygen in the atmosphere to produce carbon dioxide and water vapor. (If your burner isn't properly adjusted to give that nice blue flame, you may also get a significant amount of carbon monoxide along with carbon dioxide. This is not a good thing!) The chemical equation that represents this reaction is written like this:

$$CH_4(g) + 2\,O_2(g) \rightarrow CO_2(g) + 2\,H_2O(g)$$

You can read the equation like this: One molecule of methane gas, $CH_4(g)$, reacts with two molecules of oxygen gas, $O_2(g)$, to form one molecule of carbon dioxide gas, $CO_2(g)$, and two molecules of water vapor, $H_2O(g)$. The *2* in front of the oxygen gas and the *2* in front of the water vapor are called the reaction *coefficients.* They indicate the number of each chemical species that reacts or is formed. I show you how to figure out the value of the coefficients in the section "Balancing Chemical Reactions," later in the chapter.

Methane and oxygen (oxygen is a diatomic — two-atom — element) are the reactants, while carbon dioxide and water are the products. All the reactants and products are gases (indicated by the *g*'s in parentheses).

In this reaction, all reactants and products are invisible. The heat being evolved is the clue that tells you a reaction is taking place. By the way, this is a good example of an *exothermic* reaction, a reaction in which heat is given off. A lot of reactions are exothermic. Some reactions, however, absorb energy rather than release it. These reactions are called *endothermic* reactions. Cooking involves a lot of endothermic reactions — frying those eggs, for example. You can't just break the shells and let the eggs lie on the pan and then expect the myriad chemical reactions to take place without heating the pan (except when you're outside in Texas during August; there, the sun will heat the pan just fine).

Thinking about cooking those eggs brings to mind another issue about exothermic reactions. You have to ignite the methane coming out of the burners with a match, lighter, pilot light, or built-in electric igniter. In other words, you have to put in a little energy to get the reaction going. The energy you have to supply to get a reaction going is called the *activation energy* of the reaction. (In the next section, I show you that there's also an activation energy associated with endothermic reactions, but it isn't nearly as obvious.)

But what really happens at the molecular level when the methane and oxygen react? Divert thine eyes to the very next section to find out.

Understanding How Reactions Occur: The Collision Theory

In order for a chemical reaction to take place, the reactants must collide. It's like playing pool. In order to drop the 8-ball into the corner pocket, you must hit it with the cue ball. This collision transfers *kinetic energy* (energy of motion) from one ball to the other, sending the second ball (hopefully) toward the pocket. Energy is required to break a bond between atoms and energy is released when a bond is made. The *collision theory* states that the collision between the molecules can provide the energy needed to break the necessary bonds so that new bonds can be formed. The collision takes place at the right spot and transfers sufficient energy. The following sections provide three examples of what can happen during a collision.

Eyeing a one-step collision example

When you play pool, not every shot you make causes a ball to go into the pocket. Sometimes you don't hit the ball hard enough, and you don't transfer enough energy to get the ball to the pocket. This situation also occurs with molecular collisions and reactions. Sometimes, even if a collision takes place, not enough kinetic energy is available to be transferred — the molecules aren't moving fast enough. You can help the situation somewhat by heating the mixture of reactants. The temperature is a measure of the average kinetic energy of the molecules; raising the temperature increases the kinetic energy available to break bonds during collisions.

Sometimes, even if you hit the ball hard enough, it doesn't go into the pocket because you didn't hit it in the right spot. The same is true during a molecular collision. The molecules must collide in the right orientation, or hit at the right spot, in order for the reaction to occur.

Suppose you have an equation showing molecule *A-B* reacting with *C* to form *C-A* and *B*, like this:

A-B + C → C-A + B

The way this equation is written, the reaction requires that reactant C collide with A-B on the A end of the molecule. (You know this because the product side shows C hooked up with A — C-A.) If it hits the B end, nothing happens. The A end of this hypothetical molecule is called the *reactive site*, the place on the molecule that the collision must take place in order for the reaction to occur. If C collides at the A end of the molecule, then enough energy may be transferred to break the A-B bond. After the A-B bond is broken, the C-A bond can be formed. The equation for this reaction process can be shown in this way (I show the breaking of the A-B bond and the forming of the C-A bond as "squiggly" bonds):

   C~A~B → C-A + B

So in order for this reaction to occur, a collision between C and A-B must occur at the reactive site. The collision between C and A-B has to transfer enough energy to break the A-B bond, allowing the C-A bond to form.

If instead of having a simple A-B molecule, you have a large complex molecule, like a protein or a polymer, then the likelihood of C colliding at the reactive site is much smaller. You may have a lot of collisions, but few at the reactive site. This reaction will probably be much slower than the simple case.

Note that this example is a simple one. I've assumed that only one collision is needed, making this a one-step reaction. Many reactions are one-step, but many others require several steps before the reactants become the final products. In the process, several compounds may be formed that react with each other to give the final products. These compounds are called *intermediates*. They're shown in the reaction *mechanism*, the series of steps that the reaction goes through in going from reactants to products. But in this chapter, I keep it simple and pretty much limit my discussion to one-step reactions.

Considering an exothermic example

Imagine that the hypothetical reaction A-B + C → C-A + B is *exothermic* — a reaction in which thermal energy is given off (released) when going from reactants to products. The reactants start off at a higher energy state than the products, so energy is released in going from reactants to products. Figure 7-1 shows an energy diagram of this reaction.

In Figure 7-1, E_a is the activation energy for the reaction — the energy that you have to put in to get the reaction going. I show the collision of C and A-B with the breaking of the A-B bond and the forming of the C-A bond at the top of an activation energy hill. This grouping of reactants at the top of the activation energy hill is sometimes called the *transition state* of the reaction. As I show in Figure 7-1, the difference in the energy level of the reactants and the energy level of the products is the amount of energy (heat) that is released in the reaction.

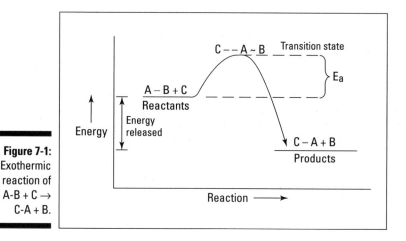

Figure 7-1:
Exothermic
reaction of
A-B + C →
C-A + B.

Some reactions may give off energy but not thermal energy. An example is light sticks. You mix two chemical solutions by flexing the light stick, and the resulting product glows — it gives off light but not heat. Another example is fireflies, which mix two chemicals in their bodies and give off light. (I remember catching fireflies in a jar for a nightlight many evenings in North Carolina. Ah, the good old days!) These reactions that give off energy are *exergonic*. If that energy is in the form of heat, the reaction is subclassified as exothermic.

Looking at an endothermic example

Suppose that the hypothetical reaction A-B + C → C-A + B is *endothermic* — a reaction in which heat is absorbed — so the reactants are at a lower energy state than the products. Figure 7-2 shows an energy diagram of this reaction.

Just as in the exothermic-reaction energy diagram shown in Figure 7-1, this diagram shows that an activation energy is associated with the reaction (represented by E_a). In going from reactants to products, you have to put in more energy initially to get the reaction started, and then you get some of that energy back out as the reaction proceeds. Notice that the transition state appears at the top of the activation energy hill — just like in the exothermic-reaction energy diagram. But although both endothermic and exothermic reactions require activation energy, exothermic reactions release thermal energy, and endothermic reactions absorb it.

Cooking is a great example of an endothermic reaction. That ground beef isn't going to be a delicious hamburger unless you cook it. You have to continually supply energy in order for the chemical reactions called cooking to take place. Cold packs that athletic trainers use to treat injuries are another example of an endothermic reaction. Two solutions in the pack are mixed, and the pack absorbs thermal energy from the surroundings. The surroundings therefore become colder.

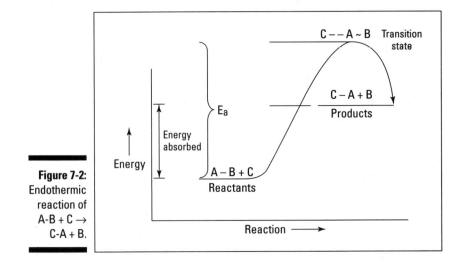

Figure 7-2:
Endothermic
reaction of
A-B + C →
C-A + B.

Other reactions may absorb energy but not necessarily heat. For example, some reactions absorb light energy in order to react. The general term that chemists use to describe reactions that absorb energy (heat or otherwise) is *endergonic*. Endothermic reactions are a subset of endergonic reactions.

Identifying Different Types of Reactions

Several general types of chemical reactions can occur based on what happens when going from reactants to products. The more common reactions are

- ✔ Combination
- ✔ Decomposition
- ✔ Single displacement
- ✔ Double displacement
- ✔ Combustion
- ✔ Redox

The following sections provide more insight into these types of reactions.

Combination reactions

In *combination reactions,* two or more reactants form one product. The reaction of sodium and chlorine to form sodium chloride,

$$2\,Na(s) + Cl_2(g) \rightarrow 2\,NaCl(s)$$

and the burning of coal (carbon) to give carbon dioxide,

$$C(s) + O_2(g) \rightarrow CO_2(g)$$

are examples of combination reactions.

Note that, depending on conditions or the relative amounts of the reactants, more than one product can be formed in a combination reaction. Take the burning of coal, for example. If an excess of oxygen is present, the product is carbon dioxide. But if a limited amount of oxygen is available, the product is carbon monoxide:

$$2\,C(s) + O_2(g) \rightarrow 2\,CO(g) \qquad \text{(limited oxygen)}$$

Decomposition reactions

Decomposition reactions are really the opposite of combination reactions. In *decomposition reactions,* a single compound breaks down into two or more simpler substances (elements and/or compounds). The decomposition of water into hydrogen and oxygen gases,

$$2\,H_2O(l) \rightarrow 2\,H_2(g) + O_2(g)$$

and the decomposition of hydrogen peroxide to form oxygen gas and water,

$$2\,H_2O_2(l) \rightarrow 2\,H_2O(l) + O_2(g)$$

are examples of decomposition reactions.

Single-displacement reactions

In *single-displacement reactions,* a more active element displaces (kicks out) another less active element from a compound. For example, if you put a piece of zinc metal into a copper(II) sulfate solution (by the way, Chapter 13 explains why copper(II) sulfate is named the way it is — in case you're wondering), the zinc displaces the copper, as shown in this equation:

$$Zn(s) + CuSO_4(aq) \rightarrow ZnSO_4(aq) + Cu(s)$$

The notation *(aq)* indicates that the compound is dissolved in water — in an *aq*ueous solution. Because zinc replaces copper in this case, it's said to be more active. If you place a piece of copper in a zinc sulfate solution, nothing happens. But how do you know which metal is the most active? An activity table lists the activity of metals. Table 7-1 shows the activity series of some common metals. Notice that because zinc is more active in the table, it will replace copper, just as the preceding equation shows.

Table 7-1	**The Activity Series of Some Common Metals**
Activity	*Metal*
Most active	Alkali and alkaline earth metals
	Al
	Zn
	Cr
	Fe
	Ni
	Sn
	Pb
	Cu
	Ag
Least active	Au

Take another look at the reaction between zinc metal and copper(II) sulfate solution:

$$Zn(s) + CuSO_4(aq) \rightarrow ZnSO_4(aq) + Cu(s) \text{ (molecular equation)}$$

I've written this reaction as a molecular equation, showing all species in the molecular/atomic form. However, these reactions normally occur in an aqueous (water) solution. When the ionically bonded $CuSO_4$ is dissolved in water, it breaks apart into *ions* (atoms or groups of atoms that have an electrical charge due to the loss or gain of electrons). The copper ion has a +2 charge because it lost two electrons. It's a *cation,* a positively charged ion. The sulfate ion has a −2 charge because it has two extra electrons. It's an *anion,* a negatively charged ion. (Check out Chapter 13 for a more complete discussion of ionic bonding.)

To show the reaction and production of ions in a reaction, you can write an *ionic equation,* like the following:

$$Zn(s) + Cu^{2+}(aq) + SO_4^{2-}(aq) \rightarrow Zn^{2+}(aq) + SO_4^{2-}(aq) + Cu(s)$$
(ionic equation)

Notice that the sulfate ion, SO_4^{2-}, doesn't change in the reaction. Ions that don't change during the reaction and are found on both sides of the equation in an identical form are called *spectator ions.* Chemists (a lazy, lazy lot, they are) often omit the spectator ions and write the equation showing only those chemical substances that are changed during the reaction. This format is called the *net-ionic equation:*

$$Zn(s) + Cu^{2+}(aq) \rightarrow Zn^{2+}(aq) + Cu(s) \qquad \text{(net-ionic equation)}$$

Double-displacement reactions

In single-displacement reactions, only one chemical species is displaced. In *double-displacement reactions,* or *metathesis reactions,* two species (normally ions) are displaced. Most of the time, reactions of this type occur in a solution, and either an insoluble solid (precipitation reactions) or water (neutralization reactions) is formed.

Precipitation reactions

If you mix a solution of potassium chloride and a solution of silver nitrate, a white insoluble solid is formed in the resulting solution. The formation of an insoluble solid in a solution is called *precipitation.* Here are the molecular, ionic, and net-ionic equations for this double-displacement reaction:

$$KCl(aq) + AgNO_3(aq) \rightarrow AgCl(s) + KNO_3(aq) \qquad \text{(molecular equation)}$$

$$K^+(aq) + Cl^-(aq) + Ag^+(aq) + NO_3^-(aq) \rightarrow AgCl(s) + K^+(aq) + NO_3^-(aq)$$
(ionic equation)

$$Cl^-(aq) + Ag^+(aq) \rightarrow AgCl(s) \qquad \text{(net-ionic equation)}$$

The white insoluble solid that's formed is silver chloride. You can drop out the potassium cation and nitrate anion spectator ions, because they don't change during the reaction and are found on both sides of the equation in an identical form. (If you're totally confused about all those plus and minus symbols in the equations, or don't know what a cation or an anion is, just flip to Chapter 13. It tells you all you need to know about this stuff.)

In order to write these equations, you have to know something about the solubility of ionic compounds. Don't fret. Here you go: If a compound is soluble, it will remain in its free ion form, but if it's insoluble, it will precipitate (form a solid). Table 7-2 gives the solubility of selected ionic compounds.

Table 7-2	Solubility of Selected Ionic Compounds
Water Soluble	**Water Insoluble**
All chlorides, bromides, iodides	Except those of Ag^+, Pb^{2+}, Hg_2^{2+}
All compounds of NH_4^+	Oxides
All compounds of alkali metals	Sulfides
All acetates	Most phosphates
All nitrates	Most hydroxides
All chlorates	
All sulfates	Except $PbSO_4$, $BaSO_4$ and $SrSO_4$

To use Table 7-2, take the cation of one reactant and combine it with the anion of the other reactant, and vice versa (keeping the neutrality of the compounds). This allows you to predict the possible products of the reaction. Then look up the solubility of the possible products in the table. If the compound is insoluble, it precipitates. If it is soluble, it remains in solution.

Neutralization reactions

The other type of double-displacement reaction is the reaction between an acid and a base. This double-displacement reaction, called a *neutralization reaction,* forms water. Take a look at the mixing solutions of sulfuric acid (auto battery acid, H_2SO_4) and sodium hydroxide (lye, NaOH). Here are the molecular, ionic, and net-ionic equations for this reaction:

$H_2SO_4(aq) + 2\ NaOH(aq) \rightarrow Na_2SO_4(aq) + 2\ H_2O(l)$
(molecular equation)

$2\ H^+(aq) + SO_4^{2-}(aq) + 2\ Na^+(aq) + 2\ OH^-(aq) \rightarrow 2\ Na^+(aq) + SO_4^{2-}(aq) + 2\ H_2O(l)$
(ionic equation)

$2 H^+(aq) + 2 OH^-(aq) \rightarrow 2 H_2O(l)$ or $H^+(aq) + OH^-(aq) \rightarrow H_2O(l)$
(net-ionic equation)

To go from the ionic equation to the net-ionic equation, the spectator ions (those that don't really react and that appear in an unchanged form on both sides on the arrow) are dropped out. Then the coefficients in front of the reactants and products are reduced down to the lowest common denominator.

You can find more about acid-base reactions in Chapter 11.

Combustion reactions

Combustion reactions occur when a compound, usually one containing carbon, combines with the oxygen gas in the air. This process is commonly called *burning.* Heat is the most useful product of most combustion reactions.

Here's the molecular equation that represents the burning of propane:

$C_3H_8(g) + 5 O_2(g) \rightarrow 3 CO_2(g) + 4 H_2O(l)$

Propane belongs to a class of compounds called *hydrocarbons,* compounds composed only of carbon and hydrogen. The product of this reaction is heat. You don't burn propane in your gas grill to add carbon dioxide to the atmosphere — you want the heat for cooking your steaks.

Combustion reactions are also a type of redox reaction, which I discuss in the following section.

Redox reactions

Redox reactions, or *reduction-oxidation reactions,* are reactions in which electrons are exchanged:

$2 Na(s) + Cl_2(g) \rightarrow 2 NaCl(s)$

$C(s) + O_2(g) \rightarrow CO_2(g)$

$Zn(s) + CuSO_4(aq) \rightarrow ZnSO_4(aq) + Cu(s)$

The preceding reactions are examples of other types of reactions (such as combination, combustion, and single-replacement reactions), but they're all also redox reactions. They all involve the transfer of electrons from one chemical species to another. Redox reactions are involved in combustion, rusting, photosynthesis, respiration, batteries, and more.

Balancing Chemical Reactions

If you carry out a chemical reaction, carefully sum up the masses of all the reactants, and compare that sum to the sum of the masses of all the products, you see that they're the same. In fact, a law in chemistry, the *law of conservation of mass,* states, "In an ordinary chemical reaction, matter is neither created nor destroyed." This means that you have neither gained nor lost any atoms during the reaction. They may be combined differently, but they're still there.

A chemical equation represents the reaction. That chemical equation is used to calculate how much of each element is needed and how much of each element will be produced. And that chemical equation must obey the law of conservation of mass.

You need to have the same number of each kind of element on both sides of the equation. The equation should balance. In this section, I show you how to balance chemical equations.

Balancing ammonia production

My favorite reaction is called the *Haber process,* a method for preparing ammonia (NH_3) by reacting nitrogen gas with hydrogen gas. This reaction is my favorite because it helps to feed the world. The ammonia that's produced is used to produce ammonium nitrate and ammonium phosphate, both of which are synthetic fertilizers that have allowed the increased production of food. Here's the equation:

$$N_2(g) + H_2(g) \rightarrow NH_3(g)$$

This equation shows you what happens in the reaction, but it doesn't show you how much of each element you need to produce the ammonia. To find out how much of each element you need, you have to balance the equation — make sure that the number of atoms on the left side of the equation equals the number of atoms on the right.

You know the reactants and the product for this reaction, and you can't change them. You can't change the compounds, and you can't change the subscripts, because that would change the compounds. So the only thing you can do to balance the equation is add *coefficients,* whole numbers in front of the compounds or elements in the equation. Coefficients tell you how many atoms or molecules you have.

For example, if you write *2 H₂O*, it means you have two water molecules:

$$2\,H_2O = H_2O + H_2O$$

Each water molecule is composed of two hydrogen atoms and one oxygen atom. So with *2 H₂O*, you have a total of 4 hydrogen atoms and 2 oxygen atoms:

$$
\begin{aligned}
2\,H_2O \quad &= H_2O + H_2O \\
&= 2\,H + 1\,O + 2\,H + 1\,O \\
&= 4\,H + 2\,O
\end{aligned}
$$

In this chapter, I show you how to balance equations by using a method called *balancing by inspection,* or as I call it, "fiddling with coefficients." You take each atom in turn and balance it by adding appropriate coefficients to one side or the other.

With that in mind, take another look at the equation for preparing ammonia:

$$N_2(g) + H_2(g) \rightarrow NH_3(g)$$

In most cases, waiting until the end to balance hydrogen atoms or oxygen atoms is a good idea based on many years of experience; balance the other atoms first.

So in this example, you need to balance the nitrogen atoms first. You have 2 nitrogen atoms on the left side of the arrow (reactant side) and only 1 nitrogen atom on the right side (product side). In order to balance the nitrogen atoms, use a coefficient of 2 in front of the ammonia on the right.

$$N_2(g) + H_2(g) \rightarrow \mathbf{2}\,NH_3(g)$$

Now you have 2 nitrogen atoms on the left and 2 nitrogen atoms on the right.

Next, tackle the hydrogen atoms. You have 2 hydrogen atoms on the left and 6 hydrogen atoms on the right (2 NH₃ molecules, each with 3 hydrogen atoms, for a total of 6 hydrogen atoms). So put a 3 in front of the H₂ on the left, giving you:

$$N_2(g) + \mathbf{3}\,H_2(g) \rightarrow 2\,NH_3(g)$$

That should do it. Do a check to be sure: You have 2 nitrogen atoms on the left and 2 nitrogen atoms on the right. You have 6 hydrogen atoms on the left ($3 \times 2 = 6$) and 6 hydrogen atoms on the right ($2 \times 3 = 6$). The equation is balanced. You can read the equation this way: 1 nitrogen molecule reacts with 3 hydrogen molecules to yield 2 ammonia molecules.

This equation also balances with coefficients of 2, 6, and 4 instead of 1, 3, and 2. In fact, any multiple of 1, 3, and 2 balances the equation, but chemists always show the lowest whole-number ratio (see the discussion on empirical formulas in Chapter 14 for details).

Flicking the lighter

In this section, follow the steps to balance the equation showing the burning of butane, a hydrocarbon, with excess oxygen available. (This is the reaction that takes place when you light a butane lighter.)

1. **Start with the unbalanced reaction written in a molecular equation.**

   The equation of burning butane with excess oxygen available is

   $$C_4H_{10}(g) + O_2(g) \rightarrow CO_2(g) + H_2O(g)$$

2. **Balance the carbon atoms first.**

   You want to wait until the end to balance hydrogen and oxygen atoms.

   You have 4 carbon atoms on the left and 1 carbon atom on the right, so add a coefficient of 4 in front of the carbon dioxide:

   $$C_4H_{10}(g) + O_2(g) \rightarrow \mathbf{4}\ CO_2(g) + H_2O(g)$$

3. **When all non-hydrogen and non-oxygen atoms are balanced, balance the hydrogen atoms.**

   Carbon is the only other atom in this example, so you can move on to hydrogen now. You have 10 hydrogen atoms on the left and 2 hydrogen atoms on the right, so use a coefficient of 5 in front of the water on the right:

   $$C_4H_{10}(g) + O_2(g) \rightarrow 4\ CO_2(g) + \mathbf{5}\ H_2O(g)$$

4. **Balance the oxygen atoms.**

   You have 2 oxygen atoms on the left and a total of 13 oxygen atoms on the right $[(4 \times 2) + (5 \times 1) = 13]$. What can you multiply 2 with in order for it to equal 13? How about 6.5?

   $$C_4H_{10}(g) + \mathbf{6.5}\ O_2(g) \rightarrow 4\ CO_2(g) + 5\ H_2O(g)$$

5. **Multiple all coefficients in the equation to get the lowest *whole-number* ratio of coefficients.**

   For this example, multiply the entire equation by 2 (just the coefficients, please) in order to generate whole numbers:

   $$[C_4H_{10}(g) + 6.5\ O_2(g) \rightarrow 4\ CO_2(g) + 5\ H_2O(g)] \times 2$$

   Multiplying every coefficient by 2 (don't touch the subscripts!) gives:

   $$2\ C_4H_{10}(g) + 13\ O_2(g) \rightarrow 8\ CO_2(g) + 10\ H_2O(g)$$

6. **Check the atom count on both sides of the equation to ensure that the equation is balanced and the coefficients are in the lowest whole-number ratio.**

Most simple reactions can be balanced in this fashion. But one class of reactions is so complex that this method doesn't work well when applied to them. They're redox reactions. A special method is used for balancing these equations, but I'll save that for another time (or another book).

Chapter 8

The Mole: Can You Dig It?

*O*ne of the many things chemists do is make new substances, a process called *synthesis*. And a logical question they ask is "How much?"

"How much of this reactant do I need to make this much product?" "How much product can I make with this much reactant?" In order to answer these questions, chemists must be able to take a balanced chemical equation, expressed in terms of atoms and molecules, and convert it to grams or pounds or tons — some type of unit they can actually weigh out in the lab. The mole concept enables chemists to move from the microscopic world of atoms and molecules to the real world of grams and kilograms and is one of the most important central concepts in chemistry. In this chapter, I introduce you to Mr. Mole.

Counting Particles by Massing

Suppose that you have a job packing 1,000 nuts and 1,000 bolts in big bags, and you get paid for each bag you fill. So what's the most efficient and quickest way of counting out nuts and bolts? Determine the mass of a hundred, or even ten, of each, and then figure out the mass of a thousand of each. Fill up the bag with nuts until it equals the mass you figured for 1,000 nuts. After you have the correct amount of nuts, use the same process to fill the bag with bolts. In other words, count by massing; that's one of the most efficient ways of counting large numbers of objects.

In chemistry, you count very large numbers of particles, such as atoms and molecules. To count them efficiently and quickly, you need to use the count-by-massing method, which means you need to know the masses of individual atoms and molecules. You can get the mass of the individual atoms on the periodic table, but what about the mass of the compounds? Well, you can simply add together the masses of the individual atoms in the compound to figure the molecular mass or formula mass. (*Molecular masses* refer to covalently bonded compounds, and *formula masses* refer to both ionic and covalent compounds. Check out Chapters 13 and 14 for details on ionic and covalent bonds.)

Water, H_2O, is composed of two hydrogen atoms and one oxygen atom. By looking on the periodic table, you can find that one hydrogen atom equals 1.0079 amu and one oxygen atom weighs 15.999 amu (*amu* stands for *atomic mass units* — see Chapter 4 for details). To calculate the molecular mass of water, you simply add together the atomic masses of two hydrogen atoms and one oxygen atom:

2×1.0079 amu = 2.016 amu	(two hydrogen atoms)
1×15.999 amu = 15.999 amu	(one oxygen atom)
2.016 amu + 15.999 amu = 18.015 amu	(the mass of a water molecule)

Now try a little harder one. Calculate the formula mass of aluminum sulfate, $Al_2(SO_4)_3$. In this salt, you have 2 aluminum atoms, 3 sulfur atoms, and 12 oxygen atoms. After you find the individual masses of the atoms on the periodic table, you can calculate the formula mass like this:

$(2 \text{ Al} \times 26.982 \text{ amu/Al}) + (3 \text{ S} \times 32.066 \text{ amu/S}) + (12 \text{ O} \times 15.999 \text{ amu/O})$

$= 342.150$ amu for $Al_2(SO_4)_3$

Using Moles to Count

When people deal with objects, they often think in terms of a convenient amount. For example, when a woman buys earrings, she normally buys a pair of them. When a man goes to the grocery store, he buys eggs by the dozen. And when I go the office supply store, I buy copy paper by the ream.

People use words to represent numbers all the time — a pair is 2, a dozen is 12, and a ream is 500. All these words are units of measure, and they're convenient for the objects they're used to measure. Rarely would you buy a ream of earrings or a pair of paper.

Likewise, when chemists deal with atoms and molecules, they need a convenient unit that takes into consideration the very small size of atoms and molecules. The *mole* is just such a unit. The following sections dive deeper into how to count with a mole.

Looking up Avogadro's number: Not in the phone book

The word *mole* stands for a number — 6.022×10^{23}. It's commonly called *Avogadro's number,* named for Amedeo Avogadro, the scientist who laid the groundwork for the mole principle.

Now a mole — 6.022×10^{23} — is a really *big* number. When written in longhand notation, it's

602,200,000,000,000,000,000,000

And *that* is why chemists like scientific notation.

If you had a mole of marshmallows, it would cover the United States to a depth of about 600 miles. A mole of rice grains would cover the land area of the world to a depth of about 75 meters. And a mole of moles . . . no, I don't even want to think about that!

Avogadro's number stands for a certain number of *things*. Normally, those things are atoms and molecules. So the mole relates to the microscopic world of atoms and molecules. But how does it relate to the macroscopic world where I work?

The answer is that a *mole* (abbreviated as *mol*) is also the number of particles in exactly 12 grams of a particular isotope of carbon (C-12). So if you have exactly 12 grams of ^{12}C, you have 6.022×10^{23} carbon atoms, which is also a mole of ^{12}C atoms. For any other element, a mole is the atomic mass expressed in grams. And for a compound, a mole is the formula (or molecular) mass in grams.

Putting moles to work

The weight of a water molecule is 18.015 amu (see the section "Counting Particles by Massing" for how to calculate the mass of compounds). Because a mole is the formula (or molecular) mass in grams of a compound, you can now say that the mass of a mole of water is 18.015 grams. You can also say that 18.015 grams of water contains 6.022×10^{23} H_2O molecules, or a mole of water. And the mole of water is composed of two moles of hydrogen and one mole of oxygen.

The mole is the bridge between the microscopic and macroscopic worlds.

6.022×10^{23} particles $\leftrightarrow$ mole $\leftrightarrow$ atomic/formula/molecular mass in grams

If you have any one of the three things — particles, moles, or grams — then you can calculate the other two.

Suppose you want to know how many water molecules are in 5.50 moles of water. You can set up a problem like this:

$$\frac{5.50 \text{ mol}}{1} \times \frac{6.022 \times 10^{23} \text{ molecules}}{1 \text{ mol}} = 3.31 \times 10^{24} \text{ molecules}$$

Or suppose that you want to know how many moles are in 25.0 grams of water. You can set up the problem like this:

$$\frac{25.0 \text{ g } H_2O}{1} \times \frac{1 \text{ mol } H_2O}{18.015 \text{ g } H_2O} = 1.39 \text{ moles } H_2O$$

You can even go from grams to particles by going through the mole.

For example, how many molecules are there in 100.0 grams of carbon dioxide?

To solve this question, do the following:

1. **Determine the molecular mass of CO_2 by looking at the periodic table.**

    The periodic table shows that one carbon atom equals 12.011 amu and one oxygen atom weighs 15.999 amu.

2. **Figure the molecular mass.**

    For this example, your calculations look like this:

    [(1 C $\times$ 12.011 g/mol C) + (2 O $\times$ 15.999 g/mol O)] = 44.01 g/mol CO_2

3. **Work the problem and solve.**

    $$\frac{100.0 \text{ g } CO_2}{1} \times \frac{1 \text{ mol } CO_2}{44.01 \text{ g}} \times \frac{6.022 \times 10^{23} \text{ molecules}}{1 \text{ mol}} = 1.368 \times 10^{24} \ CO_2 \text{ molecules}$$

And it's just as easy to go from particles to moles to grams.

Calculating empirical formulas

You can use the mole concept to calculate the empirical formula of a compound using the *percentage composition* data for that compound — the percentage by mass of each element in the compound. (The *empirical formula* indicates the different types of elements in a molecule and the lowest whole-number ration of each kind of atom in the molecule. See Chapter 14 for details.)

When I try to determine the empirical formula of a compound, I often have percentage data available. The determination of the percentage composition is one of the first analyses that a chemist does in learning about a new compound. For example, suppose I determine that a particular compound has the following weight percentage of elements present:

26.4% Na

36.8% S

36.8% O

Because I'm dealing with percentage data (amount per hundred), I assume that I have 100 grams of the compound so that my percentages can be used as masses. I then convert each mass to moles, like this:

$$\frac{26.4 \text{ g Na}}{1} \times \frac{1 \text{ mol Na}}{22.99 \text{ g}} = 1.15 \text{ mol Na}$$

$$\frac{36.8 \text{ g S}}{1} \times \frac{1 \text{ mol S}}{32.07 \text{ g}} = 1.15 \text{ mol S}$$

$$\frac{36.8 \text{ g O}}{1} \times \frac{1 \text{ mol O}}{16.00 \text{ g}} = 2.30 \text{ mol O}$$

Now I can write an empirical formula:

$Na_{1.15}S_{1.15}O_{2.30}$

I know that my subscripts have to be whole numbers, so I divide each of these by the smallest common factor, 1.15:

$$\frac{1.15 \text{ Na}}{1.15} = 1 \text{ Na} \quad \frac{1.15 \text{ S}}{1.15} = 1 \text{ S} \quad \frac{2.30 \text{ O}}{1.15} = 2 \text{ O}$$

I end up with an empirical formula of $NaSO_2$. (If a subscript is 1, it's not shown.)

I can then calculate a mass for the empirical formula by adding together the atomic masses on the periodic table of 1 sodium (Na), 1 sulfur (S), and 2 oxygen (O). The sum gives me an empirical formula mass of 87.056 grams.

Suppose, however, that in another experiment I determined the actual molecular mass of this compound to be 174.1112 grams. By dividing 174.1112 grams by 87.056 grams (actual molecular mass by the empirical formula mass), I get 2. That number means the molecular formula is twice the empirical formula, so the compound is actually $Na_2S_2O_4$.

Understanding the Role of Moles in Chemical Reactions

I think one of the reasons I enjoy being a chemist is that I like to cook. I see a lot of similarities between cooking and chemistry (except that chemists shouldn't lick the spoon). A chemist takes certain things called *reactants* and makes something new from them. A cook does the same thing. He or she takes certain things called ingredients and makes something new from them. And both the chemist and the cook follow a set of directions: The cook follows a recipe while the chemist follows the balanced chemical reaction. And both can scale up or down to produce more or less product. In the following sections I show you how.

Making the calculations

If you know the formula weight of the reactants and product, you can calculate how much you need and how much you'll get. For example, check out the *Haber reaction,* which is a method of preparing ammonia (NH_3) by reacting nitrogen gas with hydrogen gas:

$N_2(g) + 3 H_2(g) \rightarrow 2 NH_3(g)$

1 mole + 3 moles → 2 moles

All you need to do is figure the molecular weights of each reactant and product and then incorporate the weights into the equation. Use the periodic table to find the weights of the atoms and the compound (see the section "Counting Particles by Massing," earlier in this chapter, for the directions) and multiply those numbers by the number of moles, like this:

$$N_2(g) + 3\,H_2(g) \rightarrow 2\,NH_3(g)$$

1 mole + 3 moles → 2 moles

1 mol × 28.014 g/mol + 3 mol × 2.016 g/mol → 2 mol × 17.031 g/mol

In Chapter 7, I use this reaction for various examples (like I said, it's my *favorite* reaction) and explain that you can read the reaction like this: 1 molecule of nitrogen gas reacts with 3 molecules of hydrogen gas to yield 2 molecules of ammonia.

$$N_2(g) + 3\,H_2(g) \rightarrow 2\,NH_3(g)$$

1 molecule + 3 molecules → 2 molecules

Now you can scale everything up by a factor of 12:

$$N_2(g) + 3\,H_2(g) \rightarrow 2\,NH_3(g)$$

1 dozen molecules + 3 dozen molecules → 2 dozen molecules

You can even scale it up by 1,000:

$$N_2(g) + 3\,H_2(g) \rightarrow 2\,NH_3(g)$$

1,000 molecules + 3,000 molecules → 2,000 molecules

Or how about a factor of 6.022×10^{23}:

$$N_2(g) + 3\,H_2(g) \rightarrow 2\,NH_3(g)$$

6.022×10^{23} molecules + $3(6.022 \times 10^{23}$ molecules) → $2(6.022 \times 10^{23}$ molecules)

Wait a minute! Isn't 6.022×10^{23} a mole? So you can write the equation like this:

$$N_2(g) + 3\,H_2(g) \rightarrow 2\,NH_3(g)$$

1 mole + 3 moles → 2 moles

That's right — not only can those coefficients in the balanced chemical equation represent atoms and molecules, but they can also represent the number of moles.

Suppose you want to figure out how many grams of ammonia you can produce if you react 60.0 g of hydrogen gas with excess nitrogen.

Remember that the balanced chemical equation gives the mole relationship of the reactants and products, and if you have moles you have grams.

1. **Take the grams of hydrogen gas (recall that it is diatomic) and convert it to moles using its molar mass (mass of 1 mole):**

$$\frac{60.0 \text{ g H}_2}{1} \times \frac{1 \text{ mol H}_2}{2.016 \text{ g}}$$

2. **Convert the mole of hydrogen gas to moles of ammonia, using the coefficients in the balanced chemical equation:**

$$\frac{60.0 \text{ g H}_2}{1} \times \frac{1 \text{ mol H}_2}{2.016 \text{ g}} \times \frac{2 \text{ mol NH}_3}{3 \text{ mol H}_2}$$

3. **Convert from moles of ammonia to grams of ammonia using the molar mass and solve the equation.**

$$\frac{60.0 \text{ g H}_2}{1} \times \frac{1 \text{ mol H}_2}{2.016 \text{ g}} \times \frac{2 \text{ mol NH}_3}{3 \text{ mol H}_2} \times \frac{17.031 \text{ g NH}_3}{1 \text{ mol}} = 338 \text{ g NH}_3$$

Determining what you need and what you'll get: Reaction stoichiometry

After you have the weight relationships in place, you can do some stoichiometry problems. *Stoichiometry* refers to the mass relationship in balanced chemical equations.

Look at my favorite reaction — you guessed it — the Haber process:

$$N_2(g) + 3 H_2(g) \rightarrow 2 NH_3(g)$$

Suppose that you want to know how many grams of diatomic hydrogen gas will be needed to fully react with 75.00 grams of nitrogen.

The mole concept is the key. The coefficients in the balanced equation are not only the number of individual atoms or molecules but also the number of moles.

$$N_2(g) + 3 H_2(g) \rightarrow 2 NH_3(g)$$

1 mole + 3 moles $\rightarrow$ 2 moles

1. **Convert the 75.00 grams of nitrogen to moles of nitrogen.**

$$\frac{75.00 \text{ g N}_2}{1} \times \frac{1 \text{ mol N}_2}{28.014 \text{ g}}$$

2. **Use the ratio of the moles of hydrogen to the moles of nitrogen from the balanced equation to convert to moles of hydrogen.**

$$\frac{75.00 \text{ g N}_2}{1} \times \frac{1 \text{ mol N}_2}{28.014 \text{ g}} \times \frac{3 \text{ mol H}_2}{1 \text{ mol N}_2}$$

3. **Take the moles of hydrogen and convert to grams, and then solve the equation.**

$$\frac{75.00 \text{ g N}_2}{1} \times \frac{1 \text{ mol N}_2}{28.014 \text{ g}} \times \frac{3 \text{ mol H}_2}{1 \text{ mol N}_2} \times \frac{2.016 \text{ g H}_2}{1 \text{ mol}} = 16.19 \text{ g H}_2$$

The ratio of the moles H_2/mol N_2 is called a *stoichiometric ratio*. This ratio enables you to convert from the moles of one substance in a balanced chemical equation to the moles of another substance.

By the way, can you figure out how much ammonia is produced? One way is to construct a set-up similar to the preceding one, but convert from moles of nitrogen to moles of ammonia and then to grams of ammonia. The stoichiometric ratio will be different (along with the molar mass used). Another way is to reason that 75.00 grams of nitrogen reacted with 16.19 grams of hydrogen to produce 91.19 grams of ammonia (75.00 g + 16.19 g = 91.19 g).

Getting tired of the Haber process? (Me? *Never.*) Take a look at another reaction — the reduction of rust (Fe_2O_3) to iron metal by treatment with carbon (coke — a fuel, not the drink). The balanced chemical reaction looks like this:

$$2 \text{ Fe}_2\text{O}_3(s) + 3 \text{ C}(s) \rightarrow 4 \text{ Fe}(s) + 3 \text{ CO}_2(g)$$

When you get ready to work stoichiometry problems, you *must* start with a balanced chemical equation. If you don't have it to start with, you've got to go ahead and balance the equation.

In this example, the formula weights you need are

- ✔ **Fe_2O_3:** 159.70 g/mol
- ✔ **C:** 12.01 g/mol
- ✔ **Fe:** 55.85 g/mol
- ✔ **CO_2:** 44.01 g/mol

Suppose that you want to know how many grams of carbon it takes to react with 1.000 kilogram of rust.

You need to convert the kilogram of rust to grams and convert the grams to moles of rust. Then you can use a stoichiometric ratio to convert from moles of rust to moles of carbon and finally to grams. The equation looks like this:

$$\frac{1.000 \text{ kg Fe}_2\text{O}_3}{1} \times \frac{1{,}000 \text{ g}}{1 \text{ kg}} \times \frac{1 \text{ mol Fe}_2\text{O}_3}{159.70 \text{ g}} \times \frac{3 \text{ mol C}}{2 \text{ mol Fe}_2\text{O}_3} \times \frac{12.01 \text{ g C}}{1 \text{ mol C}} = 112.8 \text{ g C}$$

You can even calculate the number of carbon atoms it takes to react with that 1.000 kilogram of rust. Basically, you use the same conversions, but instead of converting from moles of carbon to grams, you convert from moles of carbon atoms using Avogadro's number:

$$\frac{1.000 \text{ kg Fe}_2\text{O}_3}{1} \times \frac{1{,}000 \text{ g}}{1 \text{ kg}} \times \frac{1 \text{ mol Fe}_2\text{O}_3}{159.70 \text{ g}} \times \frac{3 \text{ mol C}}{2 \text{ mol Fe}_2\text{O}_3} \times \frac{6.022 \times 10^{23} \text{ atoms}}{1 \text{ mol C}}$$

$$= 5.656 \times 10^{24} \text{ atoms of C}$$

Now I want to show you how to calculate the number of grams of iron produced from reacting 1.000 kilogram of rust with excess carbon. It's the same basic process as before — kilograms of rust to grams of rust to moles of rust to moles of iron to grams of iron:

$$\frac{1.000 \text{ kg Fe}_2\text{O}_3}{1} \times \frac{1{,}000 \text{ g}}{1 \text{ kg}} \times \frac{1 \text{ mol Fe}_2\text{O}_3}{159.70 \text{ g}} \times \frac{4 \text{ mol Fe}}{2 \text{ mol Fe}_2\text{O}_3} \times \frac{55.85 \text{ g Fe}}{1 \text{ mol Fe}} =$$

$$699.5 \text{ g Fe}$$

So you predict that you'll get 699.5 grams of iron metal formed. What if, however, you carry out this reaction and only get 525.0 grams of iron metal formed? Several reasons may cause you to produce less than you expect, such as sloppy technique or impure reactants. It may also be quite likely that the reaction is an equilibrium reaction, and you'll never get 100 percent conversion from reactant to products. (Equilibrium reactions don't use up all the reactants — topic for a new book.) Wouldn't it be nice if you had a way to label the efficiency of a particular reaction? You do. It's called the percent yield.

Figuring out the bang for your buck: Percent yield

In almost any reaction, you're going to produce less than expected. You may produce less because most reactions are equilibrium reactions or because some other conditions come into play. Chemists can get an idea of the efficiency of a reaction by calculating the *percent yield* for the reaction using this equation:

$$\% \text{ yield} = \frac{\text{actual yield}}{\text{theoretical yield}} \times 100\%$$

The *actual yield* is how much of the product you get when you carry out the reaction. The *theoretical yield* is how much of the product you *calculate* you'll get. The ratio of these two yields gives you an idea about how efficient the reaction is. For the reaction of rust to iron (see preceding section), your theoretical yield is 699.5 grams of iron; your actual yield is 525.0 grams. Therefore, the percent yield is

$$\% \text{ yield} = \frac{525.0 \text{ g}}{699.5 \text{ g}} \times 100\% = 75.05\%$$

A percent yield of about 75 percent isn't too bad, but chemist and chemical engineers would rather see 90+ percent. One plant using the Haber reaction has a percent yield of better than 99 percent. Now that's efficiency!

Running out of something and leaving something behind: Limiting reactants

I love to cook, and I'm always hungry. So I want to talk about making some ham sandwiches. Because I'm a chemist, I can write an equation for a ham sandwich lunch:

2 pieces of bread + 1 ham slice + 1 cheese slice → 1 ham sandwich

Suppose I check my supplies and find that I have 12 pieces of bread, 5 slices of ham, and 10 slices of cheese. How many sandwiches can I make? I can make 5, of course. I have enough bread for 6 sandwiches, enough ham for 5, and enough cheese for 10. But I'm going to run out of ham first — I'll have bread and cheese left over. And the ingredient I run out of first really limits the amount of product (sandwiches) I'll be able to make; it can be called the *limiting ingredient.*

The same is true of chemical reactions. Normally, you run out of one of the reactants and have some others left over. (In some of the problems sprinkled throughout this chapter, I tell you which reactant is the limiting one by saying you have *an excess* of the other reactant(s).)

In this section, I show you how to calculate which reactant is the limiting reactant.

Here is a reaction between ammonia and oxygen:

$$4 \text{ NH}_3(g) + 5 \text{ O}_2(g) \rightarrow 4 \text{ NO}(g) + 6 \text{ H}_2\text{O}(l)$$

Suppose that you start out with 100.0 grams of both ammonia and oxygen, and you want to know how many grams of NO (nitrogen monoxide, sometimes called nitric oxide) you can produce.

You must determine the limiting reactant and base your stoichiometric calculations on it.

A sure way to know that you have a limiting reactant problem is when you're given amounts of more than one reactant. In order to figure out which reactant is the limiting reactant, you can calculate the mole-to-coefficient ratio: You can calculate the number of moles of both ammonia and oxygen, and then you divide each by their coefficient in the balanced chemical equation. The one with the smallest mole-to-coefficient ratio is the limiting reactant. For the reaction of ammonia to nitric oxide, you can calculate the mole-to-coefficient ratio for the ammonia and oxygen like this:

$$\frac{100.0 \text{ g NH}_3}{1} \times \frac{1 \text{ mol NH}_3}{17.03 \text{ g}} = 5.87 \text{ mol} \div 4 = 1.47$$

$$\frac{100.0 \text{ g O}_2}{1} \times \frac{1 \text{ mol O}_2}{32.00 \text{ g}} = 3.13 \text{ mol} \div 5 = 0.625$$

Ammonia has a mole-to-coefficient ratio of 1.47, and oxygen has a ratio of 0.625. Because oxygen has the lowest ratio, oxygen is the limiting reactant and will determine how much NO(g) will be produced. You'll need to base your calculations on the limiting reactant, the oxygen. Now it's just another stoichiometry problem. You convert grams of oxygen to moles, to moles of NO, and to grams NO:

$$\frac{100.0 \text{ g O}_2}{1} \times \frac{1 \text{ mol O}_2}{32.00 \text{ g}} \times \frac{4 \text{ mol NO}}{5 \text{ mol O}_2} \times \frac{30.01 \text{ g NO}}{1 \text{ mol NO}} = 75.03 \text{ g NO}$$

That 75.03 grams NO is your theoretical yield. But you can even calculate the amount of ammonia left over. With the following equation, you can figure the amount of ammonia consumed:

$$\frac{100.0 \text{ g O}_2}{1} \times \frac{1 \text{ mol O}_2}{32.00 \text{ g}} \times \frac{4 \text{ mol NH}_3}{5 \text{ mol O}_2} \times \frac{17.03 \text{ g NH}_3}{1 \text{ mol NH}_3} = 42.58 \text{ g NH}_3$$

In this equation, convert from grams of oxygen gas to moles of oxygen gas and then to moles of ammonia and finally to grams of ammonia.

You started with 100.0 grams of ammonia, and you used 42.58 grams of it. The difference (100 grams – 42.58 grams = 57.42 grams) is the amount of ammonia left over.

Chapter 9

Mixing Matter Up: Solutions

*Y*ou encounter solutions all the time in everyday life. The air you breathe is a solution. That sports drink you use to replenish your electrolytes is a solution. That soft drink *and* that hard drink are both solutions. Your tap water is most likely a solution, too. In this chapter, I show you some of the properties of solutions. I introduce you to the different ways chemists represent a solution's concentration, and I tell you about the *colligative* properties of solutions (the properties of solutions that depend only on the number of particles present, not their type) and relate them to ice-cream making and antifreeze. So sit back, sip on your solution of choice, and read all about solutions.

Getting Your Definitions Straight: Solutes, Solvents, and Solutions

A *solution* is a homogeneous mixture, meaning that its properties are the same throughout. If you dissolve sugar in water and mix it really well, for example, your mixture is basically the same no matter where you sample it.

A solution is composed of a solvent and one or more solutes. The *solvent* is the substance that doesn't change state and is present in the largest amount, and the *solute* is the substance that changes state and is present in the lesser amount. You can determine which is which based on the quantities most of the time, but in a few cases of extremely soluble salts, such as lithium chloride, more than 5 grams of salt can be dissolved in 5 milliliters of water. However, water is still considered the solvent, because it's the species that has not changed state.

A solution can have more than one solute. If you dissolve salt in water to make a brine solution and then dissolve some sugar in the same solution, you have two solutes, salt and sugar. You still have only one solvent, though — water.

When I talk about solutions, most people think of liquids. But there can also be solutions of gases. Our atmosphere, for example, is a solution. Because air is almost 79 percent nitrogen, nitrogen is considered the solvent, and the oxygen, carbon dioxide, and other gases are considered the solutes. Solids can also make solutions. Alloys, for example, are solutions of one metal in another metal. Brass is a solution of zinc in copper.

Discussing solubility: How much solute will dissolve

Why do some things dissolve in one solvent and not in another? For example, oil and water don't mix to form a solution, but oil dissolves in gasoline. A general rule of solubility says that *like-dissolves-like* in regards to polarity of both the solvent and solutes. Water, for example, is a polar material; it's composed of polar covalent bonds with the positive and negative ends of the molecule. (For a rousing discussion of water and its polar covalent bonds, see Chapter 14.) Water dissolves polar solutes, such as salts and alcohols. Oil, however, is composed of largely nonpolar bonds. So water doesn't act as a suitable solvent for oil.

You know from your own experiences, I'm sure, that only so much solute can be dissolved in a given amount of solvent. You probably have been guilty of putting far too much sugar in iced tea. No matter how much you stir, there's some undissolved sugar at the bottom of the glass. The reason is that the sugar has reached its maximum solubility in water at that temperature. *Solubility* is the maximum amount of solute that will dissolve in a given amount of a solvent at a specified temperature. Solubility normally has the units of grams solute per 100 milliliters of solvent (g/100 mL).

The solubility is related to the temperature of the solvent. For solids dissolving in liquids, solubility normally increases with increasing temperature. So if you heat that iced tea, the sugar at the bottom readily dissolves. However, for gases dissolving in liquids, such as oxygen dissolving in lake water, the solubility goes down as the temperature increases. This is the basis of *thermal pollution,* the addition of heat to water that decreases the solubility of the oxygen and affects the aquatic life.

Exploring saturation

A *saturated* solution contains the maximum amount of dissolved solute possible at a given temperature. If the solution has less than the maximum amount, it's called an *unsaturated* solution. Sometimes, under unusual circumstances, the solvent may actually dissolve more than its maximum amount and become *supersaturated.* This supersaturated solution is unstable, though, and sooner or later solute will precipitate (form a solid) until the saturation point has been reached.

If a solution is unsaturated, then the amount of solute that is dissolved may vary over a wide range. A couple of rather indefinite terms describe the relative amount of solute and solvent that you can use:

✔ You can say that the solution is *dilute,* meaning that, relatively speaking, it contains very little solute per given amount of solvent. If you dissolve 0.01 grams of sodium chloride in a liter of water, for example, the solution is dilute. I once asked some students to give me an example of a dilute solution, and one replied "A $1 margarita." She was right — a lot of solvent (water) and a very little solute (tequila) are used in her example.

✔ A solution may be *concentrated,* containing a large amount of solute per the given amount of solvent. If you dissolve 200 grams of sodium chloride in a liter of water, for example, the solution is concentrated.

But suppose you dissolve 25 grams or 50 grams of sodium chloride in a liter of water? Is the solution dilute or concentrated? These terms don't hold up very well for most cases. And consider the case of IV solutions — they *must* have a very precise amount of solute in them, or the patient is in danger. So you must have a quantitative method to describe the relative amount of solute and solvent in a solution. Such a method exists — solution concentration units.

Focusing on Solution Concentration Units

You can use a variety of solution concentration units to quantitatively describe the relative amounts of the solute(s) and the solvent. In everyday life, percentage is commonly used. In chemistry, *molarity* (the moles of solute per liter of solution) is the solution concentration unit of choice. In certain circumstances, though, another unit, *molality* (the moles of solute per kilogram of solvent), is used. And I use parts per million or parts per billion when I discuss pollution control. The following sections cover some of these concentration units.

Percent composition: Three different ratios

You probably have looked at a bottle of vinegar and seen "5% acetic acid," a bottle of hydrogen peroxide and seen "3% hydrogen peroxide," or a bottle of bleach and seen "5% sodium hypochlorite." Those percentages are expressing the concentration of that particular solute in each solution. *Percentage* is the amount per one hundred. Depending on the way you choose to express the percentage, the units of amount per one hundred vary. Three different percentages are commonly used:

✔ Mass/mass (m/m) percentage

✔ Mass/volume (m/v) percentage

✔ Volume/volume (v/v) percentage

Unfortunately, although the percentage of solute is often listed, the method (m/m, m/v, v/v) is not. I normally assume that the method is weight/weight, but I'm sure you know about assumptions.

In the following sections I give some examples of solutions with these percentages. Most of the solutions are *aqueous,* solutions in which water is the solvent.

Mass/mass percentage

In *mass/mass percentage,* or *mass percentage,* the mass of the solute is divided by the mass of the solution and then multiplied by 100 to get the percentage. Another term that is sometimes used is *mass percentage (mass/mass percentage).* Normally the weight unit is grams. Mathematically, it looks like this:

$$m/m \% = \frac{\text{grams solute}}{\text{grams solution}} \times 100\%$$

If, for example, you dissolve 5.0 grams of sodium chloride in 50 grams of water, the mass percent is

$$m/m \% = \frac{5.0 \text{ g NaCl}}{55 \text{ g water}} \times 100\% = 9.1\%$$

Therefore, the solution is a 9.1 percent (m/m) solution.

Suppose that you want to make 350.0 grams of a 5 percent (w/w) sucrose, or table sugar, solution. You know that 5 percent of the weight of the solution is sugar, so you can multiply the 350.0 grams by 0.05 to get the weight of the sugar:

350.0 grams × 0.05 = 17.5 grams of sugar

The rest of the solution (350.0 grams − 17.5 grams = 332.5 grams) is water. You can simply weigh out 17.5 grams of sugar and add it to 332.5 grams of water to get your 5 percent (w/w) solution.

Mass percentage is the easiest percentage solution to make, but sometimes you may need to know the volume of the solution. In this case, you can use the mass/volume percentage.

Mass/volume percentage

Mass/volume percentage is very similar to mass/mass percentage, but instead of using grams of solution in the denominator, it uses milliliters of solution:

$$\text{m}/\text{v}\% = \frac{\text{grams solute}}{\text{mL solution}} \times 100\%$$

Suppose that you want to make 100 milliliters of a 15 percent (m/v) potassium nitrate solution.

Because you're making 100 milliliters, you already know that you're going to weigh out 15 grams of potassium nitrate (commonly called saltpeter — KNO_3). Now, here comes something that's a little different: You dissolve the 15 grams of KNO_3 in a little bit of water and dilute it to exactly 100 milliliters in a volumetric flask. In other words, you dissolve and dilute 15 grams of KNO_3 to 100 milliliters. (I tend to abbreviate *dissolve* and *dilute* by writing *d & d,* but sometimes it gets confused with Dungeons & Dragons. Yes, chemists are really, *really* nerdy.) You won't know exactly how much water you put in, but it's not important as long as the final volume is 100 milliliters.

You can also use the percentage and volume to calculate the grams of solute present. You may want to know how many grams of sodium hypochlorite are in 500 milliliters of a 5 percent (m/v) solution of household bleach. Because it is a 5 percent (m/v) solution, it has 5 grams of potassium nitrate per 100 milliliters of solution. You can set up the problem like this:

$$\frac{5 \text{ g NaOCl}}{100 \text{ mL solution}} \times \frac{500 \text{ mL solution}}{1} = 25 \text{ g NaOCl}$$

You now know that you have 25 grams of sodium hypochlorite in the 500 milliliters of solution. Sometimes both the solute and solvent are liquids. In this case, using a volume/volume percentage is more convenient.

Volume/volume percentage

With *volume/volume percentages,* both the solute and solution are expressed in milliliters:

$$\text{v}/\text{v}\% = \frac{\text{mL solute}}{\text{mL solution}} \times 100\%$$

Proof reading

When it comes to ethyl alcohol solutions, another concentration unit, called proof, is commonly used to measure the relative amount of alcohol and water. The proof is simply twice the percentage. A 50 percent ethyl alcohol solution is 100 proof. Pure ethyl alcohol (100 percent) is 200 proof. This term dates back to earlier times, when the production of ethyl alcohol for human consumption was a cottage industry. (In the part of North Carolina where I grew up, it still is a cottage industry.) There was no quality control back then, so the buyer had to be sure that the alcohol he was buying was concentrated enough (or "strong" enough) for the desired purpose. Some of the alcohol solution was poured over gunpowder and then lit. If enough alcohol was present, the gunpowder would ignite, giving "proof" that the solution was strong enough.

Ethyl alcohol (drinking alcohol) solutions are commonly made using volume/volume percentages. If you want to make 100 milliliters of a 50 percent ethyl alcohol solution, you take 50 milliliters of ethyl alcohol and dilute it to 100 milliliters with water. Again, it's a case of dissolving and diluting to the required volume. You can't simply add 50 milliliters of alcohol to 50 milliliters of water — you'd get less than 100 milliliters of solution. The polar water molecules attract the polar alcohol molecules, which tends to fill in the open framework of water molecules and prevents the volumes from simply being added together.

Molarity: It's number one!

Molarity is the concentration unit most often used by chemists, because it utilizes moles. The mole concept is central to chemistry, and molarity lets chemists easily work solutions into reaction stoichiometry. (If you're cussing me out right now because you have no idea what burrowing mammals have to do with chemistry, let alone what stoichiometry is, just flip to Chapter 8 for the scoop. Your mother would probably recommend washing your mouth out with soap first.)

Molarity (M) is defined as the moles of solute per liter of solution. Mathematically, it looks like this:

$$M = \frac{\text{mol solute}}{\text{L solution}}$$

For example, you can take 1 mole (abbreviated as *mol*) of KCl (formula mass of 74.55 g/mol — you can get the scoop on formula and molecular masses in Chapter 8, too) and dissolve and dilute the 74.55 grams to 1 liter of solution in a volumetric flask. You then have a 1-molar solution of KCl. You can label that solution as 1 M KCl. You don't add the 74.55 grams to 1 liter of water. You want to end up with a final volume of 1 liter. When preparing molar solutions, always dissolve and dilute to the required volume. This process is shown in Figure 9-1.

If 25.0 grams of KCl are dissolved and diluted to 350.0 milliliters, how would you calculate the molarity of the solution?

You know that molarity is moles of solute per liter of solution. To solve this problem, start by taking the grams and converting them to moles using the formula weight of KCl (74.55 g/mol). Divide the moles by 0.350 liters (350.0 milliliters). You can set up the equation like this:

$$\frac{25.0 \text{ g KCl}}{1} \times \frac{1 \text{ mol KCl}}{74.55 \text{ g}} \times \frac{1}{0.350 \text{ L}} = 0.958 \text{ M}$$

Now suppose that you want to prepare 2.00 liters of a 0.550 M KCl solution.

1. **Calculate how much KCl you need to weigh.**

   The solution must have 0.550 moles of potassium chloride per liter. If you convert from moles of KCl to grams of KCl, you get the number of grams of KCl per liter. You are preparing 2.00 liters, so multiplying by 2.00 liters gives you the number of grams of KCl.

   $$\frac{0.550 \text{ mol KCl}}{1 \text{ L}} \times \frac{74.55 \text{ g KCl}}{1 \text{ mol}} \times \frac{2.00 \text{ L}}{1} = 82.0 \text{ g KCl}$$

2. **Take that 82.0 grams of KCl and dissolve and dilute it to 2.00 liters.**

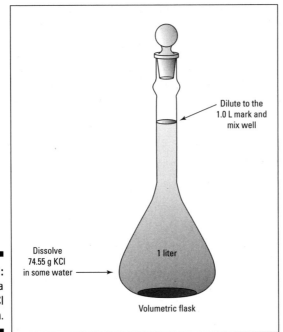

Figure 9-1:
Making a
1-molar KCl
solution.

Dissolve
74.55 g KCl
in some water

1 liter

Dilute to the
1.0 L mark and
mix well

Volumetric flask

You have another option for preparing solutions — the dilution of a more concentrated solution to a less concentrated one. For example, you can buy hydrochloric acid from the manufacturer as a concentrated solution of 12.0 M.

Suppose you want to prepare 500 milliliters of 2.0 M HCl. You can dilute some of the 12.0 M to 2.0 M, but how much of the 12.0 M HCl is needed?

To figure out the volume you need, follow these simple steps:

1. **Use the following formula:**

   $$V_{old} \times M_{old} = V_{new} \times M_{new}$$

   In the preceding equation, V_{old} is the old volume, or the volume of the original solution, M_{old} is the molarity of the original solution, V_{new} is the volume of the new solution, and M_{new} is the molarity of the new solution.

2. **Substitute the values in the equation.**

   $$V_{old} \times 12.0 \text{ M} = 500.0 \text{ mL} \times 2.0 \text{ M}$$

   $$V_{old} = (500.0 \text{ mL} \times 2.0 \text{ M})/12.0 \text{ M} = 83.3 \text{ milliliters}$$

3. **Take 83.3 milliliters of the 12.0 M HCl solution and dilute it to exactly 500.0 milliliters.**

Put about 400 milliliters of water into a 500.0 mL volumetric flask, slowly add the 83.3 milliliters of the concentrated HCl as you stir, and then dilute to the final 500 milliliters with water.

If you're actually doing a dilution of concentrated acids, be sure to *add the acid to the water* instead of the other way around! If the water is added to the concentrated acid, then so much heat will be generated that the solution will quite likely splatter all over you.

The usefulness of the molarity concentration unit is readily apparent when dealing with reaction stoichiometry.

Suppose that you want to know how many milliliters of 2.50 M sulfuric acid it takes to neutralize a solution containing 100.0 grams of sodium hydroxide.

1. **Write the balanced chemical equation for the reaction:**

$$H_2SO_4(aq) + 2\ NaOH(aq) \rightarrow 2\ H_2O(l) + Na_2SO_4(aq)$$

2. **Convert the weight of sodium hydroxide to moles (using the formula weight of NaOH, 40.00 g/mol).**

$$\frac{100.0\ g\ NaOH}{1} \times \frac{1\ mol\ NaOH}{40.00\ g\ NaOH}$$

3. **Convert from moles of NaOH to moles of H_2SO_4.**

$$\frac{100.0\ g\ NaOH}{1} \times \frac{1\ mol\ NaOH}{40.00\ g\ NaOH} \times \frac{1\ mol\ H_2SO_4}{2\ mol\ NaOH}$$

4. **Calculate the volume using the molarity of the acid solution and convert the volume to milliliters.**

$$\frac{100.0\ g\ NaOH}{1} \times \frac{1\ mol\ NaOH}{40.00\ g\ NaOH} \times \frac{1\ mol\ H_2SO_4}{2\ mol\ NaOH} \times \frac{1\ L}{2.50\ mol\ H_2SO_4} \times$$

$$\frac{1{,}000\ mL}{1\ L} = 500.0\ mL$$

500.0 milliliters of the 2.50 M H_2SO_4 solution is needed to completely react with the solution that contains 100.0 grams of NaOH.

Molality: Another use for the mole

Molality is another concentration term that involves moles of solute. It isn't used very much except in dealing with colligative properties, but I want to tell you a little about it, just in case you happen to run across it. *Molality (m)* is defined as the moles of solute per kilogram of solvent. It's one of the few concentration units that doesn't use the solution's weight or volume. Mathematically, it looks like this:

$$m = \frac{mol\ solute}{kg\ solvent}$$

Suppose you want to dissolve 15.0 grams of NaCl in 50.0 grams of water.

To calculate the molality, you need to convert the 50.0 grams of water to kilograms (0.0500 kg). Then convert the grams of NaCl to moles of NaCl and divide by the kilograms of water, as in the following equation:

$$\frac{15.0 \text{ g NaCl}}{1} \times \frac{1 \text{ mol NaCl}}{58.44 \text{ g NaCl}} \times \frac{1}{0.0500 \text{ kg}} = 5.13 \text{ m}$$

Parts per million: The pollution unit

Percentage and molarity, and even molality, are convenient units for the solutions that chemists routinely make in the lab or the solutions that are commonly found in nature. However, if you begin to examine the concentrations of certain pollutants in the environment, you find that those concentrations are very, very small. Percentage and molarity work when you're measuring very dilute solutions found in the environment, but they're not very convenient. In order to express the concentrations of very dilute solutions, scientists have developed another concentration unit — parts per million.

Percentage is parts per hundred, or grams solute per 100 grams of solution. *Parts per million (ppm)* is grams solute per one million grams of solution. It's most commonly expressed as milligrams solute per kilogram solution, which is the same ratio. The reason it's expressed this way is that chemists can easily weigh out milligrams or even tenths of milligrams, and, if you're talking about aqueous solutions, a kilogram of solution is the same as a liter of solution. (The density of water is 1 gram per milliliter, or 1 kilogram per liter. The weight of the solute in these solutions is so very small that it's negligible when converting from the mass of the solution to the volume.)

By law, the maximum contamination level of lead in drinking water is 0.05 ppm. This number corresponds to 0.05 milligrams of lead per liter of water. That's pretty dilute. But mercury is regulated at the 0.002 ppm level. Sometimes, even this unit isn't sensitive enough, so environmentalists have resorted to the parts per billion (ppb) or parts per trillion (ppt) concentration units. Some neurotoxins are deadly at the parts per billion level.

Comprehending Colligative Properties of Solutions

Some properties of solutions depend on the specific nature of the solute. In other words, an effect you can record about the solution depends on the specific identity of the solute. For example, salt solutions taste salty, whereas sugar solutions taste sweet. Salt solutions conduct electricity (they're electrolytes — see Chapter 13), but sugar solutions don't (they're nonelectrolytes). Solutions containing the nickel cation are commonly green, and those containing the copper cation are blue.

Some properties of solution don't depend on the specific type of solute — just the *number* of solute particles. Properties that simply depend on the relative number of solute particles are called *colligative properties*. The effect you can record about the solution depends on the number of solute particles present. The following sections discuss in greater detail these colligative properties — these effects — including

- ✔ Vapor-pressure lowering
- ✔ Boiling-point elevation
- ✔ Freezing-point depression
- ✔ Osmotic pressure

Reducing the pressure: Vapor-pressure lowering

If a liquid is contained in a closed container, the liquid eventually evaporates, and the gaseous molecules contribute to the pressure above the liquid. The pressure due to the gaseous molecules of the evaporated liquid is called the liquid's *vapor pressure.*

If you make that same liquid the solvent in a solution, the vapor pressure due to the solvent evaporation is lower because the solute particles in the liquid take up space at the surface and the solvent can't evaporate as easily. And many times, the solute and solvent may have an attraction that also makes it more difficult for the solvent to evaporate. That lowered vapor pressure is independent of what kind of solute you use. Instead, it depends on the number of solute particles.

For example, if you add one mole of sucrose to a liter of water and add one mole of dextrose to another liter of water, the amount that the pressure lowers is the same, because you're adding the same *number* of solute particles. If, however, you add a mole of sodium chloride to a liter of water, the vapor pressure lowers by twice the amount of the sucrose or glucose solutions. Sodium chloride breaks apart into two ions, so adding a mole of sodium chloride yields two moles of particles (ions), and the greater number of solute particles leads to lower pressure.

This lowering of vapor pressure partially explains why the Great Salt Lake has a lower evaporation rate than you may expect. The salt concentration is so high that the vapor pressure (and evaporation) has been significantly lowered.

Using antifreeze in summer: Boiling-point elevation

Each individual liquid has a specific temperature at which it boils (at a given atmospheric pressure). This temperature is the liquid's *boiling point.* If you use a particular liquid as a solvent in a solution, you find that the boiling point of the solution is always higher than the pure liquid. This is called the *boiling-point elevation.*

Boiling-point elevation explains why you don't replace your antifreeze with pure water in the summer. You want the coolant to boil at a higher temperature so that it absorbs as much engine heat as possible *without* boiling. You also use a pressure cap on your radiator, because the higher the pressure, the higher the boiling point. This concept also explains why a pinch of salt in the cooking water causes foods to cook a little faster. The salt raises the boiling point so that more energy can be transferred to cooking the food during a given amount of time.

You can actually calculate the amount of boiling-point elevation by using this formula:

$$\Delta T_b = K_b m$$

ΔT_b is the *increase* in the boiling point, K_b is the boiling-point elevation constant (0.512°C/m for water), and m is the molality of particles. (For molecular substances, the molality of particles is the same as the molality of the substance; for ionic compounds, you have to take into consideration the formation of ions and calculate the molality of the ion particles.) Solvents other than water have a different boiling point elevation constant (K_b).

Suppose you want to know the boiling point of 2.0 m aqueous KNO_3 solution.

KNO_3 is a salt, a strong electrolyte. In a 2.0 m solution you have 4.0 m in particles, because KNO_3 dissociates completely into K^+ and NO_3^-. Therefore:

$$\Delta T_b = K_b m$$
$$\Delta T_b = (0.512°C/m) \times (4.0\ m)$$
$$\Delta T_b = 2.0°C$$

The change in boiling point is 2.0 degrees Celsius. You know that the boiling point of a solution is always *higher* that the pure solvent, so the solution's boiling point is:

$$100.0°C + 2.0°C = 102.0°C$$

Making ice cream: Freezing-point depression

Each individual liquid has a specific temperature at which it freezes. If you use a particular liquid as a solvent in a solution, though, you find that the freezing point of the solution is always lower than the pure liquid. This is called the *freezing-point depression,* and it's a colligative property of a solution, meaning that it depends on the number of solute particles.

The depression of the freezing point of a solution relative to the pure solvent explains why you put rock salt in the ice/water mix when making homemade ice cream. The rock salt forms a solution with a lower freezing point than water (or the ice cream mix that's to be frozen). The freezing-point depression effect also explains why a salt (normally calcium chloride, $CaCl_2$) is spread on ice to melt it. The dissolving of calcium chloride is highly exothermic (it gives off a lot of heat). As the calcium chloride dissolves, it melts the ice and forms a solution in the resulting water. The salt solution that's formed when the ice melts has a lowered freezing point that keeps the solution from refreezing. Freezing-point depression also explains the use of antifreeze in your cooling system during the winter. The more you use (up to a concentration of 50/50), the lower the freezing point.

In case you're interested, you can actually calculate the amount the freezing point will be depressed:

$$\Delta T_f = K_f m$$

ΔT_f is the amount the freezing point will be lowered, K_f is the freezing point depression constant (1.86°C/m for water), and m is the molality of the particles.

Suppose you want to calculate the freezing point of that 2.0 m aqueous KNO_3 solution used in the previous section.

You know that the solution is 4.0 m in particles, so

$$\Delta T_f = K_f m$$
$$\Delta T_f = (1.86°C/m) \times 4.0\ m$$
$$\Delta T_f = 7.4°C$$

You know that the freezing point of the solution is lower that the freezing point of water (the solvent), so the freezing point of the solution is

$$0.0°C - 7.4°C = -7.4°C$$

Figure 9-2 shows the effect of a solute on both the freezing point and the boiling point of a solvent.

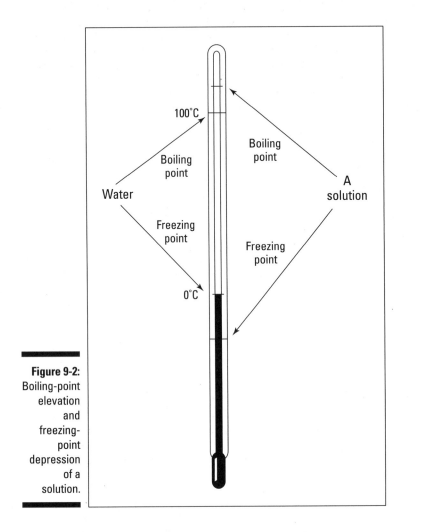

Figure 9-2:
Boiling-point elevation and freezing-point depression of a solution.

Keeping blood cells alive and well: Osmotic pressure

Suppose that you take a container and divide it into two compartments with a thin membrane containing microscopic pores large enough to allow water molecules but not solute particles to pass through. This membrane type is called a *semipermeable membrane;* it lets only some small particles pass through.

You then add a concentrated salt solution to one compartment and a more dilute salt solution to the other. Initially, the two solution levels start out the same. But after a while, you notice that the level on the more concentrated side has risen, and the level on the more dilute side has dropped. This change in levels is due to the passage of water molecules from the more dilute side to the more concentrated side through the semipermeable membrane. This process is called *osmosis,* the passage of a solvent through a semipermeable membrane into a solution of higher solute concentration. The pressure that you have to exert on the more concentrated side in order to stop this process is called *osmotic pressure.* This process is shown in Figure 9-3.

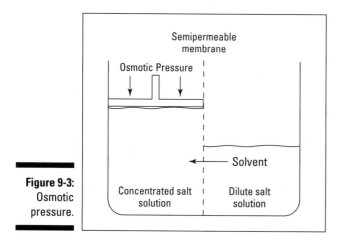

Figure 9-3: Osmotic pressure.

The solvent always flows through the semipermeable membrane from the more dilute side to the more concentrated side. In fact, you can have pure water on one side and any salt solution on the other, and water always goes from the pure-water side to the salt-solution side. The more concentrated the salt solution, the more pressure it takes to stop the osmosis (the higher the osmotic pressure).

But what if you apply more pressure than is necessary to stop the osmotic process, exceeding the osmotic pressure? Water is forced through the semipermeable membrane from the more concentrated side to the more dilute side, a process called *reverse osmosis.* Reverse osmosis is a good, relatively inexpensive way of purifying water. My local "water store" uses this process to purify drinking water (so-called "RO water"). The world has many reverse osmosis plants that extract drinking water from seawater. Navy pilots even carry small reverse osmosis units with them in case they have to eject at sea.

Cell walls often act as semipermeable membranes. Do you ever eat pickles? Cucumbers are soaked in a brine solution in order to make pickles. The concentration of the solution inside the cucumber is less than the concentration of the brine solution, so water migrates through the cell walls into the brine, causing the cucumber to shrink.

One of the most biologically important consequences of osmotic pressure involves the cells within our own bodies. You can look at red blood cells as an example. Inside the blood cell is an aqueous solution, and outside the cell is another aqueous solution (intercellular fluid). When the solution outside the cell has the same osmotic pressure as the solution inside the cell, it's said to be *isotonic*. Water can be exchanged in both directions, helping to keep the cell healthy. However, if the intercellular fluid becomes more concentrated and has a higher osmotic pressure *(hypertonic)*, water flows primarily out of the blood cell, causing it to shrink and become irregular in shape. This is a process called *crenation*. The process may occur if the person becomes seriously dehydrated, and the crenated cells are not as efficient in carrying oxygen. If, on the other hand, the intercellular fluid is more dilute than the solution inside the cells and has a lower osmotic pressure *(hypotonic)*, the water flows mostly into the cell. This process, called *hemolysis,* causes the cell to swell and eventually rupture. Figure 9-4 shows crenation and hemolysis.

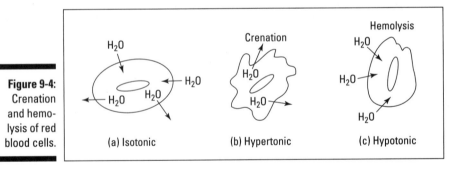

Figure 9-4:
Crenation and hemolysis of red blood cells.

(a) Isotonic   (b) Hypertonic   (c) Hypotonic

The processes of crenation and hemolysis explain why the concentration of IV solutions is so very critical. If they're too dilute, then hemolysis can take place, and if they're too concentrated, crenation is a possibility.

You can calculate the osmotic pressure (π) by using the following equation:

$$\pi = (nRT/V)i = iMRT$$

In this equation π is the osmotic pressure in atmospheres, n is the number of moles of solute, R is the ideal gas law constant (0.0821 L atm/K mol), T is the Kelvin temperature, V is the volume of the solution, and i is the van 't Hoff factor (the number of moles of particles that will be formed from 1 mole of solute). n/V may be replaced by M, the molarity of the solution.

Clearing the Air on Colloids

If you dissolve table salt in water, you form an aqueous solution. The solute particle size is very small — around 1 nanometer (nm), which is 1×10^{-9} meters. This solute doesn't settle to the bottom of a glass, and it can't be filtered out of the solution.

If, however, you go down to your local stream and dip out a glass of water, you'll notice that there's a lot of material in it. Many of the solute particles are larger than 1,000 nm. They quickly settle to the bottom of the glass and can be filtered out. In this case, you have a *suspension* and not a solution. Whether you have one or the other depends on the size of the solute particles.

But an intermediate range exists between solutions and suspensions. When the solute particle size is 1 to 1,000 nanometers, you have a *colloid.* Solutes in colloids don't settle out like they do in suspensions. In fact, distinguishing colloids from true solutions is sometimes difficult. One of the few ways to distinguish between them is to shine a light through the suspected liquid. If it's a true solution, with very small solute particles, the light beam will be invisible. If you have a colloid, however, you'll be able to see the light beam as it reflects off the relatively large solute particles. This is called the *Tyndall effect,* and it's shown in Figure 9-5.

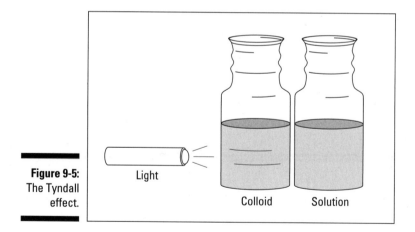

Figure 9-5:
The Tyndall effect.

Light

Colloid   Solution

Colloids come in many varieties. Have you ever eaten a marshmallow? It's a colloid of a gas in a solid. Whipped cream is a colloid of a gas in a liquid. Have you ever driven through the fog and seen your headlight beams? You were experiencing the Tyndall effect of a liquid-in-a-gas colloid. Smoke is a colloid of a solid (ash or soot) in a gas (air). Air pollution problems are often caused by the stability of this type of colloid.

Chapter 10

Thermochemistry: Hot Stuff

*R*eactants form products, but something else goes on as well during a chemical reaction — energy changes. In this chapter, I focus on one type of energy change that takes place — changes in heat.

Many chemical reactions that you may be able to mention are *exothermic reactions,* reactions that release energy. Burning natural gas to run the stove when you fry chicken, burning wood in a campfire to keep warm, and the combustion of gasoline in your automobile's engine are all example of exothermic reactions. In many cases, the energy produced is the reaction product that's desired.

In other reactions, energy is absorbed as the reaction takes place. These reactions are *endothermic,* and energy appears on the reactants' side at the start. Cooking is a great example of endothermic reactions. Crack a couple of eggs into a hot frying pan, and energy from the pan is absorbed into the eggs during the myriad chemical reactions called cooking.

Thermochemistry is the study of the energy changes that take place during a chemical reaction. In this chapter, you investigate the concepts associated with thermochemistry. You discover how to calculate the amount of heat produced during a specific chemical reaction and how much heat is needed to cause a desired reaction to take place. So turn on the air conditioning and grab a glass of cold tea — this chapter may heat things up a bit.

Looking at Reactions and Energy Changes

Thermochemistry is concerned with the heat changes that occur during a chemical reaction (or a physical change such as melting). It's part of a wider field of chemistry called *thermodynamics,* the general study of energy transfers. But before you can study these energy changes, I first have to define a few terms. The following sections give you an overview of some important concepts you need to understand before delving deeper into thermochemistry.

Systems and surroundings

In thermochemistry, thermal (heat) energy changes occur between the system and surroundings. The *system* is that part of the universe being studied (the place in which the chemical reaction takes place), and the *surroundings* are the rest of the universe that is being affected by the system. Suppose you're carrying out a reaction in a beaker in the chemistry lab. The beaker is the system, and the surroundings are the lab. Although some scientists may say the rest of the universe is the surroundings, for practical purposes, the beaker is the system. You can feel energy changes in the beaker, but I doubt that some alien millions of light years away will be able to ever detect it.

Heat

Heat (normally represented as q) is the energy that flows when a system and surroundings are at different temperatures. The energy always flows from the area of higher temperature to the area of lower temperature. If q has a positive value ($+q$), the system has absorbed energy via heat transfer from the surroundings (an *endothermic* process). If q has a negative value ($-q$), the system has transferred heat to the surroundings (an *exothermic* process).

The transferred heat is an *extensive property,* which means its amount depends on the amount of matter involved. For example, more fire means more heat, so you get more heat from a large fire than from a small fire, even if the large fire isn't hotter. *Temperature* (the average kinetic energy of matter) is an *intensive property,* a property that's not affected by the amount of matter involved. The temperature of your latte may be the same as the temperature of your bath water, but much more energy (heat) was required to warm up the bath water.

Units of energy

The two units that you use in thermochemistry calculations are the *joule* and the *calorie*.

- ✔ **Joule (J)** is the SI unit of energy. It has the units of kg m^2/s^2. Remember that 1 J = 1 kg m^2/s^2.

- ✔ **Calorie (cal)** is the amount of energy needed to raise the temperature of 1 gram of water 1 degree Celsius. By definition a calorie is exactly 4.184 J.

Although both units may be used, most older textbooks (and older teachers) tend to use calories, while newer text tend to use joules. I tend to use both, just so you're familiar with both.

I'm sure you've heard the term *calories* before. Who hasn't talked about cutting back on calories or wondering how many calories that triple bacon cheeseburger contains? But those calories are different — they're nutritional Calories (notice the capital *C*). Nutritional Calories are really kilocalories, 1,000 calories. So that 300-Calorie candy bar really contains 300,000 calories of energy that must either be used up or, more likely, stored as fat. Now that's sticker shock! Food sellers count on 300 Calories sounding better than 300,000 calories.

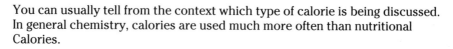

You can usually tell from the context which type of calorie is being discussed. In general chemistry, calories are used much more often than nutritional Calories.

Engineers often use the British thermal unit (BTU). A BTU was originally defined as the amount of heat required to raise the temperature of 1 pound (0.454 kg) of liquid water by 1 degree Fahrenheit (0.556°C) at a constant pressure of 1 atmosphere. It's equivalent to 1,054 joules and is commonly used when referring to the power output of steam engines and air-conditioning units.

Heat capacities

If you heat a substance, whether it's water, copper, or mercury, the temperature of the substance increases, but the magnitude of that temperature change varies from substance to substance. The *specific heat capacity, s* (sometimes simply called the *specific heat*), of a substance is the amount of heat that's required to raise the temperature of 1 gram of that substance 1 degree Celsius and can be measured in units of J/°C or cal/°C.

To calculate heat capacity, take the heat absorbed or released by a certain mass of that substance during a chemical process and divide it by the mass (in grams) times the temperature change ($/(\text{mass})\Delta T$), as in the following equation:

$$s = q/(\text{mass})\Delta T$$

The specific heats of substances vary somewhat with temperature, so you need to know the specific temperature of the substance before trying to calculate s. The temperature variance isn't very great, so you normally can use the tabulated values of specific heats that have been measured at around 25 degrees Celsius. For example, one gram of H_2O, water, absorbs 4.184 J or exactly 1 calorie when increasing its temperature from 25 to 26 degrees Celsius. It also loses that same amount of energy when cooling from 26 to 25 degrees. The units of specific heat capacity are J/g°C or cal/g°C.

Because units of Celsius and Kelvin have the same magnitude, you may use a Kelvin temperature change in place of a Celsius temperature change when calculating specific heat capacities.

If you know the specific heat of a substance, its mass, and the change in temperature, you can use the following equation to calculate the amount of heat, q, that was involved in the process:

$$q = s(\text{mass})(\Delta T)$$

An even more useful heat capacity quantity is the molar heat capacity. The *molar heat capacity, S,* is the amount of heat needed to raise the temperature of one mole of a substance one degree Celsius (or Kelvin). It has units of J/mole°C or cal/mole°C. Table 10-1 lists the specific heat capacities and molar heat capacities of some common substances.

Table 10-1   Specific and Molar Heat Capacities of Selected Substances at 25 Degrees Celsius

Substance	Specific Heat Capacity (J/g°C)	Molar Heat Capacity (J/mole°C)
Al (solid)	0.90	24.3
C (diamond)	0.50	6.0
C (graphite)	0.72	8.6
Cu (solid)	0.39	24.5
Fe (solid)	0.44	24.8
H_2O (solid)	2.09	37.7
H_2O (liquid)	4.18	75.3
H_2O (gas)	2.03	36.4

You can see in the table that liquid water has a relatively high heat capacity. This capacity allows the oceans and other bodies of water to absorb large amounts of heat, helping to keep the earth's temperature moderate. But the table also shows that water's three states of matter have different heat capacities. Be careful when pulling a heat capacity value from this or any other table that you use the heat capacity value for the correct state.

The different allotropes of an element also have different heat capacities. (*Allotropes* are different forms of an element where the atoms of the element are bonded together in different ways.) Table 10-1 includes the heat capacities of graphite and diamond, two allotropic forms of carbon.

Here is a typical thermochemical problem you may encounter that addresses the points in this section:

> How much heat does it take to raise the temperature of 50.0 grams of solid iron from 20.0 to 35.0 degrees Celsius?

Looking at Table 10-1, you see that the specific heat capacity of iron (Fe) is 0.44 J/g°C. And you know that $s = q/(mass)(\Delta T)$, so $q = s(mass)(\Delta T)$.

Substitute the given values to get this answer:

> $q = (0.44 \text{ J/g°C})(50.0 \text{ g})(35.0 - 20.0°C) = 330 \text{ J}$

Calorimetry

In order to measure the energy change that takes place during a chemical process (the study of *calorimetry*), chemists use an instrument called a *calorimeter*. The following sections take a look at the two general types of calorimeters used by chemists: constant-pressure and constant-volume.

Constant-pressure calorimeters

The constant-pressure calorimeter is the simpler of the two types. It's used to measure the energy changes taking place at a constant pressure (q_p), which is useful when studying processes involving solutions, such as in acid-base neutralization reactions and dilution, or in measuring the specific heats of materials. The calorimeter holds the pressure constant by exposing the reaction container to the atmosphere, which keeps the pressure of the reaction at atmospheric pressure.

You can make a constant-pressure calorimeter easily by stacking two Styrofoam coffee cups together. Using two stacked cups really insulates the system from the surroundings. In addition, you need to provide some way to stir the solution, and you need a thermometer to measure the temperature change. You can run the thermometer and stirrer through a lid which helps

to provide additional insulation. Figure 10-1 illustrates how a coffee cup calorimeter may be constructed.

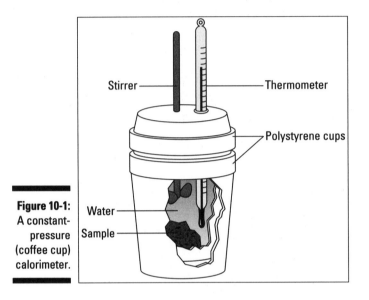

Figure 10-1:
A constant-
pressure
(coffee cup)
calorimeter.

Stirrer —————————— Thermometer

Polystyrene cups

Water —

Sample —

Styrofoam has a very low thermal conductivity, so hardly any of the heat of
the reaction in absorbed by the cups. Therefore, the heat changes that occur
are contained within the solution and recorded by the thermometer.

The heat lost or gained by the process is equal to the heat lost or gained by
the solution. Knowing the heat capacity of the solution, its mass, and the
temperature change allows you to calculate the heat of reaction (q_{rxn}).

You can use this same method to measure the heat capacity of a substance:
Heat a known mass of a substance to a certain temperature, and then add it
to the calorimeter containing a known mass of water at a known temperature.
You know that the heat lost by the substance equals the heat gained by the
water:

$$-q_{solid} = q_{water}$$

If you then substitute the mathematical relationship for the heat capacities,
you get the following equation. Because you know everything except the
specific heat of the solid, you can solve for it.

$$-(s_{solid})(\text{mass}_{solid})(\Delta T_{solid}) = (s_{water})(\text{mass}_{water})(\Delta T_{water})$$

Constant-volume (bomb) calorimeter

Constant-volume calorimeters (commonly called *bomb calorimeters*) are used to measure energy changes that occur during combustion reactions. Figure 10-2 is a diagram of a typical bomb calorimeter.

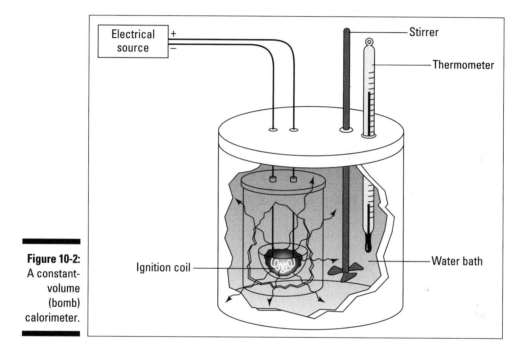

Figure 10-2: A constant-volume (bomb) calorimeter.

To use this type of calorimeter, follow these steps:

1. **Place a weighed sample of the substance to be tested into a cup and place it into a heavy-walled reaction vessel called the bomb.**

   An electrical ignition apparatus is in contact with the sample.

2. **Evacuate the air in the bomb and replace it with oxygen.**

3. **Place the bomb in an insulated container filled with a known amount of water.**

4. **Insert a thermometer in the water to measure temperature changes while a stirrer circulates the water around the bomb.**

5. **After giving everything an opportunity to stabilize, ignite the sample electrically.**

The sample burns in the oxygen gas. The energy given off by the combustion of the sample is absorbed by the water inside the calorimeter and by the calorimeter itself.

6. **Compare the temperature of the water before the combustion occurs to the temperature after combustion.**

   The heat absorbed by the water (q_{water}) is the specific heat capacity of the water (in J/g°C) times the mass of water (in g) times the change in temperature (final temperature [$°C_{final}$] minus initial temperature [$°C_{initial}$]):

   $$q_{water} = (J/g°C)(g)(°C_{final} - °C_{initial})$$

   The amount of energy absorbed by the calorimeter is equal to the heat capacity of the calorimeter times the change in temperature (final temperature – initial temperature):

   $$q_{cal} = (J/°C)(°C_{final} - °C_{initial})$$

   The heat of reaction, q_{rxn}, is equal to the energy absorbed by the water plus the energy absorbed by the calorimeter:

   $$q_{rxn} = q_{water} + q_{cal}$$

7. **Calculate the specific heat capacity or the molar heat capacity of the substance.**

   Use the equation you used in the previous section "Heat capacities" and solve for s:

   $$q = s(mass)(\Delta T)$$
   $$s = q/(mass)(\Delta T)$$

If you know the molar mass of the substance, you can then convert the grams to moles and calculate the molar heat capacity.

Understanding Enthalpy Changes

A vast majority of the reactions that chemists study are reactions at constant pressure, because the pressure can be held constant at atmospheric pressure by simply having the reaction vessel (the beaker or flask, for example) open to the atmosphere. Because these reactions are so common, the energy change at constant pressure has a special name, *enthalpy*. The *enthalpy change* (ΔH) is the heat lost or gained by the system during a chemical reaction while under constant pressure conditions. The following conventions apply to ΔH:

✔ If $\Delta H < 0$, then energy is released and the reaction is *exothermic*.

✔ If $\Delta H > 0$, then energy is absorbed and the reaction is *endothermic*.

For example, consider the reaction of hydrogen and oxygen gases to form a mole of water under constant pressure conditions. The state of gas is indicated by *(g)*.

$$H_2(g) + \frac{1}{2}O_2(g) \rightarrow H_2O(g) \quad \Delta H = -241.8 \text{ kJ}$$

Although this enthalpy change is somewhat difficult to measure under constant pressure conditions, it can be calculated in other ways, as you see later in this chapter.

The negative sign on the enthalpy indicates that this reaction is exothermic; energy is released. (I used to do this reaction as part of my chemical magic show, and I can testify that a lot of energy is released!) This energy change is many times called the enthalpy (heat) of reaction, $\Delta H_{reaction}$ or ΔH_{rxn}.

A *thermochemical equation* is a balanced chemical equation that shows not only the mole relationship between reactants and products but also the enthalpy change associated with that specific reaction. Note that in a thermochemical equation, having fractional coefficients is perfectly okay (unlike in ordinary chemical reactions), and the coefficients always refer to moles of reactants or products, not individual molecules.

Be sure to pay attention to the states of the reactants and products. You also want to remember two other conventions that apply to thermochemical equations:

✔ **Because the enthalpy change is an extensive property, if you use a multiplier on the equation, you use that same multiplier on the ΔH.**

For example, if you want to write the thermochemical equation for the production of 2 moles of water, you have to multiply the entire equation above, including the ΔH, by 2.

$$2\left[H_2(g) + \frac{1}{2}O_2(g) \rightarrow H_2O(g) \quad \Delta H = -241.8 \text{ kJ}\right]$$

$$2 H_2(g) + O_2(g) \rightarrow 2 H_2O(g) \qquad \Delta H = -483.6 \text{ kJ}$$

✔ **If you reverse a thermochemical equation, you reverse the sign on the enthalpy change.** For example, if you want to write the thermochemical equation for the decomposition of a mole of water, simply reverse the initial equation for the formation of a mole of water and change the sign of the ΔH.

$$H_2O(g) \rightarrow H_2(g) + \tfrac{1}{2}O_2(g) \Delta H = +241.8 \text{ kJ}$$

This equation tells you that if you want to decompose a mole of water (about 18 grams), you must supply 241.8 kJ of energy.

Finding Heats of Reaction

To the industrial chemist, the balanced chemical equation allows her to calculate amounts of reactants needed and amounts of products formed. A chemistry student has to determine the same kinds of calculations when running a chemical reaction in the lab (or on paper in the middle of an exam). But the energy changes that take place are just as important. When doing experiments (either in the instructional lab or the industrial lab), you want to know whether you have to continuously supply heat during the reaction or get rid of heat being produced and how much heat will need to be supplied or disposed of. The answers to these questions can be found with the thermochemical equation and the associated ΔH (enthalpy change — see the preceding section). The following sections explain the four main ways of determining heats of reactions.

Doing it yourself

You can go into the lab and determine the heats of reactions (ΔH_{rxn}) yourself. That way, you can set the conditions of temperature, pressure, and so on to exactly what you want. For uncommon reactions that aren't well documented, taking the necessary steps to determine the heats works fine, but it's a time-consuming and costly process. Wouldn't it be nice if you were able to find a tabulated value or calculate the enthalpy of reaction from known values, instead of doing all that messy lab work? In fact, most of the time you can do just that, as the next three methods show.

Referring to tables

If you're fortunate, you can find the specific reaction you're looking for in a table (maybe in the back of your chemistry textbook) and use the data provided there. Earlier in this chapter I identified and used several thermo-chemical equations, such as in the decomposition reaction for liquid water:

$$H_2O(g) \rightarrow H_2(g) + \tfrac{1}{2}O_2(g) \Delta H = +241.8 \text{ kJ}$$

Simply referring to a table is a great procedure if you can find the reaction you want, but many times this quick fix isn't possible.

Relying on Hess's law

If you can find some related reactions and their heat changes (ΔH), you have the option of a method called *Hess's law calculations*. Enthalpy is a *state function*, which means the path taken during the reaction doesn't matter; all that matters are the initial and final states. The enthalpy change is the same whether the reaction occurs in one step or in a series of steps. The basis of Hess's law is that you can calculate the enthalpy change for a desired reaction by manipulating associated reactions until you reach the desired reaction. Hess's law states that if you can write a desired reaction as the sum of two or more known reactions, then the enthalpy change for the desired reaction is simply the sum of the enthalpy changes for the known reactions.

Suppose you want to determine the enthalpy change in kJ for the following reaction, where *(s)* indicates a solid state and *(g)* means gas:

$$2\ C(s) + H_2(g) \rightarrow C_2H_2(g) \qquad \Delta H = ?$$

ΔH can be determined experimentally with a great deal of difficulty, but the easier option is to use the following known reactions:

$$(1)\,C(s) + O_2(g) \rightarrow CO_2(g)\ \Delta H = -393.5\ \text{kJ}$$

$$(2)\,H_2(g) + \frac{1}{2}O_2(g) \rightarrow H_2O(l)\ \Delta H = -285.8\ \text{kJ}$$

$$(3)\,C_2H_2(g) + \frac{5}{2}O_2(g) \rightarrow 2\,CO_2(g) + H_2O(l)\ \Delta H = -1299.8\ \text{kJ}$$

Because the coefficient in front of the carbon is 2 in the desired reaction, you multiply reaction (1) by 2. $C_2H_2(g)$ is the product in the desired reaction, so reverse reaction (3), changing the sign of its ΔH. Keep reaction (2) unchanged, because you need one H_2 on the left. If you add together these three reactions, you get the desired reaction, and simply adding together the enthalpy changes of the three reactions gives you the enthalpy change for the desired reaction:

$$2\left[C(s) + O_2(g) \rightarrow CO_2(g)\right] \Delta H = 2(-393.5\ \text{kJ})$$

$$+H_2(g) + \frac{1}{2}O_2(g) \rightarrow H_2O(l)\ \Delta H = -285.8\ \text{kJ}$$

$$+2\,CO_2 + H_2O(l) \rightarrow C_2H_2(g) + \frac{5}{2}O_2(g)\ \Delta H = +1299.8\ \text{kJ}$$

$$= 2\,C(s) + H_2(g) \rightarrow C_2H_2(g)\ \Delta H = 227.0\ \text{kJ}$$

You can save time and effort by calculating the ΔH for the reaction you want from other thermochemical reactions. But what if those other thermochemical reactions are not available? That's where the fourth method comes in, using standard heats of formation.

Using standard heats of formation

If you can't find appropriate equations, you can calculate heats of reaction using standard enthalpies of formation. Standard enthalpies of formation, ΔH_f° (where the superscript $^\circ$ indicates standard conditions and the subscript f indicates a formation reaction), are the enthalpy changes associated with the formation of one mole of a substance from its elements with all materials in their standard states. The standard state for thermodynamic properties is the state (form) in which the substance is found at 1 bar of pressure (approximately atmospheric pressure at sea level).

The standard heat of formation for an element in its standard state is exactly 0. Table 10-2 lists the standard enthalpies of formation for a number of chemical substances, and you can normally find a more complete table in the back of your chemistry text.

Table 10-2	Standard Enthalpies of Formation at 25°C (298K)
Substance	**ΔH_f°(kJ/mole)**
C (graphite)	0
C (diamond)	1.896
CH_4 (gas)	−74.87
C_2H_2 (gas)	227
C_2H_4 (gas)	52.47
CO_2 (gas)	−393.5
Fe_2O_3 (solid)	−825.5
H_2O (gas)	−241.826
H_2O (liquid)	−285.840
NH_3 (gas)	−45.9
NH_3 (aq)	−80.83
N_2O (gas)	82.05
NO (gas)	90.29
NO_2 (gas)	33.2

These tabulated standard enthalpies of formation may be used to calculate the standard enthalpy (heat) of reaction, ΔH_{rxn}, the enthalpy of reaction at standard conditions.

Calculate the standard heat of reaction by taking the sum of the standard heats of formation of all products (taking into consideration the number of moles of each) minus the standard heats of formation of all reactants (again taking into consideration the number of moles of each reactant):

$$\Delta H_{rxn} = \Sigma \Delta H_f(\text{products}) - \Sigma \Delta H_f(\text{reactants})$$

The symbol Σ means "the sum of." If a substance has more than one mole present, you must use the coefficient in the balanced chemical equation as a multiplier.

Consider the following reaction, used in the previous example:

$$2\,CO_2 + H_2O(l) \rightarrow C_2H_2(g) + \frac{5}{2}\,O_2(g)$$

Use standard enthalpies of formation to calculate the standard enthalpy of reaction. Be sure when using Table 10-2 to choose the substance in the right physical state. For example, in this problem you use the enthalpy of formation for liquid water, not gaseous water.

$$\Delta H_{rxn} = \Sigma \Delta H_f(\text{products}) - \Sigma \Delta H_f(\text{reactants})$$

$$= \left[1\,\text{mole}\left(\Delta H_f^{\circ}\left(C_2H_2(g)\right)\right) + \frac{5}{2}\,\text{moles}\left(\Delta H_f^{\circ}\left(O_2(g)\right)\right)\right] -$$
$$\left[2\,\text{moles}\left(\Delta H_f^{\circ}\left(CO_2(g)\right)\right) + 1\,\text{mole}\left(\Delta H_f^{\circ}\left(H_2O(l)\right)\right)\right]$$

$$= \left[1\,\text{mole}\left(227\frac{kJ}{mole}\right) + \frac{5}{2}\,\text{moles}\left(0\frac{kJ}{mole}\right)\right] -$$
$$\left[2\,\text{moles}\left(-393.5\frac{kJ}{mole}\right) + 1\,\text{mole}\left(-285.840\frac{kJ}{mole}\right)\right]$$

$$= 1299.8\,kJ$$

Notice that this result agrees with the enthalpy of reaction used previously in this chapter.

Uncovering Enthalpies and Phase Transitions

Any time that a substance changes phase, whether melting or boiling or subliming, the temperature remains constant, but an enthalpy (energy change at constant pressure) value is associated with that phase change. If your teacher wants you to calculate the total energy change that takes place between two temperatures that include one or more phase changes, then the following discussion should help.

For example, when ice is melting, the temperature remains constant at 0 degrees Celsius until all the ice is converted to water. The heat that is absorbed during this phase change overcomes the strong intermolecular forces in the solid ice and breaks down the crystalline structure of the water. The heat required to convert one mole of a substance from a solid at its melting point to a liquid at its melting point is the *molar enthalpy (heat) of fusion* (ΔH_{fus}). Water has a heat of fusion of 6.01 kJ/mole.

The process of going from a liquid to a solid at constant temperature, the opposite of melting (fusion), is *solidification (crystallization)*. The enthalpy change is the *molar enthalpy of solidification, ΔH_{sol}*, and is equal in magnitude but opposite in sign to the molar heat of fusion, making it –6.01 kJ/mole for water.

When water starts to boil, converting from a liquid to a gas, the same temperature stability occurs. The energy heating the water is used to overcome the intermolecular forces in the liquid, but from the time that water reaches 100 degrees Celsius as $H_2O(l)$ until it has all been converted to $H_2O(g)$, the temperature remains constant. The heat required to vaporize one mole of a substance at its boiling point is the *molar enthalpy (heat) of vaporization* (ΔH_{vap}). For water, the molar enthalpy of vaporization is 40.7 kJ/mole.

What happens if you start with steam, $H_2O(g)$, and cool it until it begins to condense? The process of going from a gas to a liquid at constant temperature, *condensation,* is the opposite of vaporization. Water's molar enthalpy of vaporization, ΔH_{vap}, is 40.7 kJ/mole. Therefore, the molar enthalpy of condensation, ΔH_{con}, is –40.7 kJ/mole for water.

Most substances change from solid to liquid to gas as temperature rises. However, some substances, such as dry ice, $CO_2(s)$, undergo *sublimation,* changing directly from the solid state to the gaseous state without becoming a liquid. *Molar enthalpy of sublimation, ΔH_{sub}*, is the enthalpy change when one mole of a solid substance sublimes. The enthalpy of sublimation is simply the sum of the molar heats of fusion and vaporization:

$$\Delta H_{sub} = \Delta H_{fus} + \Delta H_{vap}$$

Deposition, the opposite of sublimation, is going directly from the gaseous state to the solid state without first becoming a liquid. The *molar enthalpy of deposition, ΔH_{dep}*, is equal in magnitude but opposite in sign from the molar heat of sublimation.

Chapter 11

Sour and Bitter: Acids and Bases

. . . .

In This Chapter

▶ Discovering the properties of acids and bases

▶ Finding out about the two acid-base theories

▶ Determining what makes acids and bases strong

▶ Using indicators to quickly identify bases and acids

▶ Taking a look at the pH scale

▶ Figuring out buffers

. . . .

*A*cids and bases are indicators of pH, and if you walk into any kitchen or bathroom, you find a multitude of each. In the refrigerator you find soft drinks full of carbonic acid. The pantry holds vinegar and baking soda, an acid and a base. Peek under the sink, and you notice the ammonia and other cleaners, most of which are bases. Check out that can of lye-based drain opener — it's highly basic. In the medicine cabinet, you find aspirin, an acid, and antacids, bases. The everyday world is full of acids and bases, and so is the everyday world of the industrial chemist. In this chapter, I cover acids and bases and some good basic chemistry.

Getting to Know the Properties of Acids and Bases: Macroscopic View

Before you can fully grasp an appreciation for acids and bases, you first need a basic understanding of what they are. The following lists look at the properties of acids and bases that you can observe in everyday life.

Acids:

- ✔ Taste sour (but remember, in the lab, you test, not taste)
- ✔ Produce a painful sensation on the skin
- ✔ React with certain metals (magnesium, zinc, and iron) to produce hydrogen gas
- ✔ React with limestone and baking soda to produce carbon dioxide
- ✔ React with litmus paper and turn it red

Bases:

- ✔ Taste bitter (again, in the lab, you test, not taste)
- ✔ Feel slippery on the skin
- ✔ React with oils and greases
- ✔ React with acids to produce a salt and water
- ✔ React with litmus paper and turn it blue

Tables 11-1 and 11-2 show some common acids and bases found around the home.

Table 11-1	Common Acids Found in the Home	
Chemical Name	*Formula*	*Common Name or Use*
Hydrochloric acid	HCl	Muratic acid
Acetic acid	CH_3COOH	Vinegar
Sulfuric acid	H_2SO_4	Auto battery acid
Carbonic acid	H_2CO_3	Carbonated water
Boric acid	H_3BO_3	Antiseptic; eye drops
Acetylsalicylic acid	$C_{16}H_{12}O_6$	Aspirin

Table 11-2	Common Bases Found in the Home	
Chemical Name	*Formula*	*Common Name or Use*
Ammonia	NH_3	Cleaner
Sodium hydroxide	NaOH	Lye
Sodium bicarbonate	$NaHCO_3$	Baking soda
Magnesium hydroxide	$Mg(OH)_2$	Milk of magnesia

Chemical Name	Formula	Common Name or Use
Calcium carbonate	$CaCO_3$	Antacid
Aluminum hydroxide	$Al(OH)_3$	Antacid

Recognizing Acids and Bases: Microscopic View

If you look at Tables 11-1 and 11-2 closely, you may notice that all the acids contain hydrogen, and most of the bases contain the hydroxide ion (OH^-). Two main theories of the structure of acids and bases use these facts in their descriptions of acids and bases and their reactions:

- ✔ Arrhenius theory
- ✔ Bronsted-Lowry theory

The following sections take a closer look at these two theories to help you gain a firmer understanding of acids and bases.

The Arrhenius theory: Must have water

The Arrhenius theory was the first modern acid-base theory developed. In this theory, an acid is a substance that yields H^+ (hydrogen) ions when dissolved in water, and a base is a substance that yields OH^- (hydroxide) ions when dissolved in water. $HCl(g)$ can be considered a typical Arrhenius acid, because when this gas dissolves in water, it *ionizes* (forms ions) to give the H^+ ion. (Chapter 13 is where you need to go for the riveting details about ions.)

$$HCl(aq) \rightarrow H^+(aq) + Cl^-(aq)$$

According to the Arrhenius theory, sodium hydroxide is classified as a base, because when it dissolves, it yields the hydroxide ion:

$$NaOH(aq) \rightarrow Na^+(aq) + OH^-(aq)$$

Arrhenius also classified the reaction between an acid and a base as a *neutralization* reaction, because if you mix an acidic solution with a basic solution, you end up with a neutral solution composed of water and a salt.

$$HCl(aq) + NaOH(aq) \rightarrow H_2O(l) + NaCl(aq)$$

Look at the ionic form of this equation (the form showing the reaction and production of ions) to see where the water comes from:

$$H^+(aq) + Cl^-(aq) + Na^+(aq) + OH^-(aq) \rightarrow H_2O(l) + Na^+(aq) + Cl^-(aq)$$

As you can see, the water is formed from combining the hydrogen and hydroxide ions. In fact, the net-ionic equation (the equation showing only those chemical substances that are changed during the reaction) is the same for all Arrhenius acid-base reactions:

$$H^+(aq) + OH^-(aq) \rightarrow H_2O(l)$$

The Arrhenius theory is still used quite a bit. But, like all theories, it has some limitations. It specifies that the reactions must take place in water and that bases must contain hydroxide ions, but many reactions that do not meet these stipulations resemble acid-base reactions. For example, look at the gas phase reaction between ammonia and hydrogen chloride gases:

$$NH_3(g) + HCl(g) \rightarrow NH_4^+(aq) + Cl^-(aq) \rightarrow NH_4Cl(s)$$

The two clear, colorless gases mix, and a white solid of ammonium chloride forms. I show the intermediate formation of the ions in the equation so that you can better see what's actually happening. The HCl transfers one H^+ to the ammonia. That's basically the same thing that happens in the HCl/NaOH reaction, but the reaction involving the ammonia can't be classified as an acid-base reaction, because it doesn't occur in water and it doesn't involve the hydroxide ion. But again, the same basic process is taking place in both cases. In order to account for these similarities, a new acid-base theory was developed, the Bronsted-Lowry theory.

The Bronsted-Lowry acid-base theory: Giving and accepting

The Bronsted-Lowry theory attempts to overcome the limitations of the Arrhenius theory by defining an acid as a proton (H^+) donor and a base as a proton (H^+) acceptor. The base accepts the H^+ by furnishing a lone pair of electrons for a *coordinate-covalent* bond, which is a covalent bond (shared pair of electrons) in which one atom furnishes both of the electrons for the bond. Normally, one atom furnishes one electron for the bond and the other atom furnishes the second electron (see Chapter 14). In the coordinate-covalent bond, one atom furnishes both bonding electrons.

Figure 11-1 shows the NH_3/HCl reaction using the electron-dot structures of the reactants and products. (Electron-dot structures are covered in Chapter 14, too.)

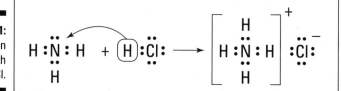

Figure 11-1:
Reaction
of NH_3 with
HCl.

HCl is the acid, so it's the proton donor, and ammonia is the base, the proton acceptor. Ammonia has a lone pair of nonbonding electrons that it can furnish for the coordinate-covalent bond.

I discuss acid-base reactions under the Bronsted-Lowry theory in the section "Competing for protons: Bronsted-Lowry acid-base reactions," later in this chapter.

Distinguishing between Strong and Weak Acids and Bases

I want to introduce you to a couple different categories of acids and bases — strong and weak. However, remember that acid-base strength is not the same as concentration. *Strength* refers to the amount of ionization or breaking apart that a particular acid or base undergoes. *Concentration* refers to the amount of acid or base that you initially have. You can have a concentrated solution of a weak acid, or a dilute solution of a strong acid, or a concentrated solution of a strong acid, or . . . well, I'm sure you get the idea. The following sections point out the main differences between strong and weak acids and bases.

Ionizing completely: Strong acids

Acids that ionize completely are considered strong. If you dissolve hydrogen chloride gas in water, the HCl reacts with the water molecules and donates a proton to them:

$$HCl(g) + H_2O(l) \rightarrow Cl^-(aq) + H_3O^+(aq)$$

The H_3O^+ ion is called the *hydronium ion*. This reaction goes essentially to completion, meaning the reactants keep creating the product until they're all used up. In this case, all the HCl ionizes to H_3O^+ and Cl^-; no more HCl is present. Note that water, in this case, acts as a base, accepting the proton from the hydrogen chloride.

Because strong acids ionize completely, calculating the concentration of the hydronium ion and chloride ion in solution is easy if you know the initial concentration of the strong acid. For example, suppose that you bubble 0.1 moles (see Chapter 8 to get a firm grip on moles) of HCl gas into a liter of water. You can say that the initial concentration of HCl is 0.1 M (0.1 mol/L). *M* stands for molarity, and *mol/L* stands for moles of solute per liter. (For a detailed discussion of molarity and other concentration units, see Chapter 9.)

You can represent this 0.1 M concentration for the HCl in this fashion: [HCl] = 0.1. The brackets around the compound indicate molar concentration, or mol/L. Because the HCl completely ionizes, you see from the balanced equation that for every HCl that ionizes, you get one hydronium ion and one chloride ion. So the concentration of ions in that 0.1 M HCl solution is

$[H_3O^+]$ = 0.1 M and $[Cl^-]$ = 0.1 M

This idea is valuable when you calculate the pH of a solution. (And you can do just that in the section "Putting Coffee and Other Substances on the pH Scale," later in this chapter.) Table 11-3 lists the most common strong acids you're likely to encounter.

Table 11-3	Common Strong Acids
Name	*Formula*
Hydrochloric acid	HCl
Hydrobromic acid	HBr
Hydroiodic acid	HI
Nitric acid	HNO_3
Perchloric acid	$HClO_4$
Sulfuric acid (first ionization only)	H_2SO_4

Sulfuric acid is called a *diprotic* acid. It can donate 2 protons, but only the first ionization goes 100 percent. The other acids listed in Table 11-3 are *monoprotic* acids, because they donate only one proton.

Falling to pieces: Strong bases

A strong base *dissociates* (breaks apart) completely in water. You normally see only one strong base, the hydroxide ion, OH^-. Calculating the hydroxide ion concentration is really straightforward. Suppose that you have a 1.5 M (1.5 mol/L) NaOH solution. The sodium hydroxide, a salt, completely dissociates into ions:

$$NaOH \rightarrow Na^+(aq) + OH^-(aq)$$

If you start with 1.5 mol/L NaOH, then you have the same concentration of ions:

$$[Na^+] = 1.5 \text{ M and } [OH^-] = 1.5 \text{ M}$$

Ionizing partway: Weak acids

Acids that only partially ionize are called *weak acids*. One example is acetic acid (CH_3COOH). If you dissolve acetic acid in water, it reacts with the water molecules, donating a proton and forming hydronium ions. It also establishes an equilibrium in which you have a significant amount of unionized acetic acid. (In reactions that go to completion, the reactants are completely used up creating the products. But in equilibrium systems, two exactly opposite chemical reactions — one on each side of the reaction arrow — are occurring at the same place, at the same time, with the same speed of reaction.)

The acetic acid reaction with water looks like this:

$$CH_3COOH(l) + H_2O(l) \rightleftarrows CH_3COO^-(aq) + H_3O^+(aq)$$

The acetic acid that you added to the water is only partially ionized, so it's a weak acid. In the case of acetic acid, about 5 percent ionizes, and 95 percent remains in the molecular form. The amount of hydronium ion that you get in solutions of acids that don't ionize completely is much less than it is with a strong acid.

 Calculating the hydronium ion concentration in weak acid solutions isn't as straightforward as it is in strong solutions, because not all the weak acid that dissolves initially has ionized. In order to calculate the hydronium ion concentration, you must use the equilibrium constant expression for the weak acid. For weak acid solutions, you use a mathematical expression called the K_a — the *acid ionization constant*.

Take a look at the generalized ionization of some weak acid HA (hypothetical acid):

$$HA + H_2O \rightleftarrows A^- + H_3O^+$$

The K_a expression for this weak acid is

$$K_a = \frac{\left[H_3O^+\right]\left[A^-\right]}{\left[HA\right]}$$

Note that the [HA] represents the molar concentration of HA *at equilibrium*, not initially. Also, note that the concentration of water doesn't appear in the K_a expression, because there's so much water that it actually becomes a constant incorporated into the K_a expression.

Now go back to that acetic acid equilibrium. The K_a for acetic acid is 1.8×10^{-5}. The K_a expression for the acetic acid ionization is

$$K_a = 1.8 \times 10^{-5} = \frac{\left[H_3O^+\right]\left[CH_3COO^-\right]}{\left[CH_3COOH\right]}$$

Use this K_a to calculate the hydronium ion concentration in a 2.0 M solution of acetic acid.

1. **Start by considering what information you have about the initial concentration and the products.**

   The initial concentration of acetic acid is 2.0 M, and you know that a little bit has ionized, forming a little hydronium ion and acetate ion. You also can see from the balanced reaction that for every hydronium ion that's formed, an acetate ion is also formed — so their concentrations are the same.

2. **Using the information you've gathered from the balanced reaction, represent the amount of [H$_3$O$^+$] and [CH$_3$COO$^-$] as x:**

   $[H_3O^+] = [CH_3COO^-] = x$

3. **In order to produce the x amount of hydronium and acetate ion, the same amount of ionizing acetic acid is required. Represent the amount of acetic acid remaining at equilibrium as the amount you started with.**

   In this example, you started with 2.0 M, minus the amount that ionizes, x:

   $[CH_3COOH] = 2.0 - x$

4. **For the vast majority of situations, you can say that x is very small in comparison to the initial concentration of the weak acid. Therefore, you can often approximate the equilibrium concentration of the weak acid with its initial concentration.**

   In this case, you can say that $2.0 - x$ is approximately equal to 2.0. The equilibrium constant expression now looks like this:

   $$K_a = 1.8 \times 10^{-5} = \frac{\left[x\right]\left[x\right]}{\left[2.0\right]} = \frac{\left[x\right]^2}{\left[2.0\right]}$$

5. **Solve for *x*, which is the [H$_3$O$^+$]:**

$$(1.8 \times 10^{-5})[2.0] = [x]^2$$

$$\sqrt{3.6 \times 10^{-5}} = [x] = \left[H_3O^+ \right]$$

$$6.0 \times 10^{-3} = [H_3O^+]$$

Refer to Table 11-3 to see some common strong acids. Most of the other acids you encounter are weak.

One way to distinguish between strong and weak acids is to look for an acid ionization constant (K$_a$) value. If the acid has a K$_a$ value, then it's weak.

Finding equilibrium with water: Weak bases

Weak bases also react with water to establish an equilibrium system. Ammonia is a typical weak base. It reacts with water to form the ammonium ion and the hydroxide ion:

$$NH_3(g) + H_2O(l) \rightleftharpoons NH_4^+ + OH^-$$

Like a weak acid, a weak base is only partially ionized. The modified equilibrium constant expression for weak bases is K$_b$. You use it exactly the same way you use the K$_a$ (see "Ionizing partway: Weak acids" for the details) except you solve for the [OH$^-$].

Competing for protons: Bronsted-Lowry acid-base reactions

With the Arrhenius theory, acid-base reactions are neutralization reactions. With the Bronsted-Lowry theory, acid-base reactions are a competition for a proton. For example, take a look at the reaction of ammonia with water:

$$NH_3(g) + H_2O(l) \rightleftharpoons NH_4^+(aq) + OH^-(aq)$$

Ammonia is a base (it accepts the proton), and water is an acid (it donates the proton) in the forward (left to right) reaction. But in the reverse reaction (right to left), the ammonium ion is an acid and the hydroxide ion is a base. If water is a stronger acid than the ammonium ion, then a relatively large concentration of ammonium and hydroxide ions are at equilibrium. If, however, the ammonium ion is a stronger acid, much more ammonia than ammonium ion is present at equilibrium.

Bronsted and Lowry said that an acid reacts with a base to form conjugate acid-base pairs, which differ by a single H^+. NH_3 is a base, for example, and NH_4^+ is its conjugate acid. H_2O is an acid in the reaction between ammonia and water, and OH^- is its conjugate base. In this reaction, the hydroxide ion is a strong base and ammonia is a weak base, so the equilibrium is shifted to the left — not much hydroxide is present at equilibrium.

Playing both parts: Amphoteric water

Water can act as either an acid or a base, depending on what it's combined with. When an acid reacts with water, water acts as a base, or a proton acceptor. But in reactions with a base (like ammonia; see the preceding section), water acts as an acid, or a proton donor. Substances that can act as either an acid or a base are called *amphoteric*.

But can water react with itself? Yes, it can. When two water molecules react with each other, one donates a proton and the other accepts it:

$$H_2O(l) + H_2O(l) \rightleftarrows H_3O^+(aq) + OH^-(aq)$$

This reaction is an *equilibrium reaction*. A modified equilibrium constant, called the K_w (which stands for *water dissociation constant*), is associated with this reaction. The K_w has a value of 1.0×10^{-14} and has the following form:

$$1.0 \times 10^{-14} = K_w = [H_3O^+][OH^-]$$

In pure water, the $[H_3O^+]$ equals the $[OH^-]$ from the balanced equation, so $[H_3O^+] = [OH^-] = 1.0 \times 10^{-7}$. The K_w value is a constant. This value allows you to convert from $[H^+]$ to $[OH^-]$, and vice versa, in *any* aqueous solution, not just pure water. In aqueous solutions, the hydronium ion and hydroxide ion concentrations are rarely going to be equal. But if you know one of them, the K_w allows you to figure out the other one.

Take a look at the 2.0 M acetic acid solution problem in the section "Ionizing partway: Weak acids," earlier in this chapter. You find that the $[H_3O^+]$ is 6.0×10^{-3}. Now you have a way to calculate the $[OH^-]$ in the solution by using the K_w relationship:

$$K_w = 1.0 \times 10^{-14} = \qquad [H_3O^+][OH^-]$$
$$1.0 \times 10^{-14} = \qquad [6.0 \times 10^{-3}][OH^-]$$
$$1.0 \times 10^{-14}/6.0 \times 10^{-3} = \ [OH^-]$$
$$1.7 \times 10^{-12} = \qquad [OH^-]$$

Identifying Acids and Bases with Indicators

Indicators are substances (organic dyes) that change color in the presence of an acid or base. You may be familiar with an acid-base indicator plant — the hydrangea. If it's grown in acidic soil, it turns pink; if it's grown in alkaline soil, it turns blue. Another common substance that acts as a good acid-base indicator is red cabbage. I have my students chop some up and boil it (most of them really *love* this part). They then use the leftover liquid to test substances. When mixed with an acid, the liquid turns pink; when mixed with a base, it turns green. In fact, if you take some of this liquid, make it slightly basic, and then exhale your breath into it through a straw, the solution eventually turns pink, indicating that the solution has turned slightly acidic. The carbon dioxide in your breath reacts with the water, forming carbonic acid:

$$CO_2(g) + H_2O(l) \rightleftarrows H_2CO_3(aq)$$

Carbonated beverages are slightly acidic due to this reaction. Carbon dioxide is injected into the liquid to give it fizz. A little of this carbon dioxide reacts with the water to form carbonic acid. This reaction also explains why rainwater is slightly acidic. It absorbs carbon dioxide from the atmosphere as it falls to earth.

In chemistry, indicators are used to indicate the presence of an acid or a base. Chemists have many indicators that change at slightly different pH levels. (You've probably heard the term *pH* used in various contexts. Me, I even remember it being used to sell deodorant and shampoo on TV. If you want to know what it actually stands for, check out the section "Putting Coffee and Other Substances on the pH Scale.") The two most commonly used indicators, which I discuss in the following sections, are

- Litmus paper
- Phenolphthalein

Taking a quick dip with litmus paper

Litmus is a substance that is extracted from a type of lichen and absorbed into porous paper. (In case you're scheduled for a hot game of Trivial Pursuit this weekend, *lichen* is a plant that's made up of an alga and a fungus that live intimately together and mutually benefit from the relationship. Sounds kind of sordid to me.)

There are three different types of litmus:

- ✔ Red litmus is used to test for bases.
- ✔ Blue litmus is used to test for acids.
- ✔ Neutral litmus can be used to test for both.

If a solution is acidic, both blue and neutral litmus turn red. If a solution is basic, both red and neutral litmus turn blue. Litmus paper is a good, quick test for acids and bases. And you don't have to put up with the smell of boiling cabbage.

Titrating with phenolphthalein

Phenolphthalein (pronounced fe-nul-*tha*-leen) is another commonly used indicator. Until a few years ago, phenolphthalein was used as the active ingredient in a popular laxative. In fact, I used to extract the phenolphthalein from the laxative by soaking it in either rubbing alcohol or gin (being careful not to drink it). I'd then use this solution as an indicator.

Phenolphthalein is clear and colorless in an acid solution and pink in a basic solution. It's commonly used in a procedure called a *titration,* where the concentration of an acid or base is determined by its reaction with a base or acid of known concentration.

Determine the molar concentration of an HCl solution.

To calculate, following these steps:

1. **Place a known volume (say, 25.00 milliliters measured accurately with a pipette) in an Erlenmeyer flask and add a couple drops of phenolphthalein solution.**

   An *Erlenmeyer flask* is a flat-bottomed, conical-shaped container.

   Because you're adding the indicator to an acidic solution, the solution in the flask remains clear and colorless.

2. **Add small amounts of a standardized sodium hydroxide solution of known molarity (for example, 0.100 M) with a buret.**

   A *buret* is a graduated glass tube with a small opening and a stopcock, which helps you measure precise volumes of a solution.

3. **Keep adding the base until the solution turns the faintest shade of pink.**

   I call this the *endpoint* of the titration, the point in which the indicator shows that the acid has been exactly neutralized by the base. Figure 11-2 shows the titration setup.

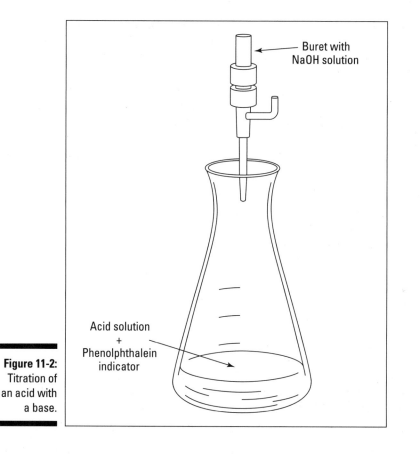

Buret with
NaOH solution

Acid solution
+
Phenolphthalein
indicator

Figure 11-2:
Titration of
an acid with
a base.

Suppose that it takes 35.50 milliliters of the 0.100 M NaOH to reach the endpoint of the titration of the 25.00 milliliters of the HCl solution. Here's the reaction:

$$HCl(aq) + NaOH(aq) \rightarrow H_2O(l) + NaCl(aq)$$

From the balanced equation, you can see that the acid and base react in a 1:1 mole ratio. So if you can calculate the moles of bases added, you'll also know the number of moles of HCl present. Knowing the volume of the acid solution then allows you to calculate the molarity (note that you convert the milliliters to liters so that your units cancel nicely):

$$\frac{0.100 \text{ mol NaOH}}{1} \times \frac{0.03550 \text{ L}}{1} \times \frac{1 \text{ mol HCl}}{1 \text{ mol NaOH}} \times \frac{1}{0.02500 \text{ L}} = 0.142 \text{ M HCl}$$

You can calculate the titration of a base with a standard acid solution (one of known concentration) in exactly the same way, except the endpoint is the first disappearance of the pink color.

Putting Coffee and Other Substances on the pH Scale

The amount of acidity in a solution is related to the concentration of the hydronium ion in the solution. The more acidic the solution is, the larger the concentration of the hydronium ion. In other words, a solution in which the $[H_3O^+]$ equals 1.0×10^{-2} is more acidic than a solution in which the $[H_3O^+]$ equals 1.0×10^{-7}. The *pH scale,* a scale based on the $[H_3O^+]$, was developed to more easily tell at a glance the relative acidity of a solution. *pH* is defined as the negative logarithm (abbreviated as *log*) of the $[H_3O^+]$. Mathematically, it looks like this:

$$pH = -\log [H_3O^+]$$

Based on the water dissociation constant, K_w (see "Playing both parts: Amphoteric water," earlier in this chapter), in pure water the $[H_3O^+]$ equals 1.0×10^{-7}. Using this mathematical relationship, you can calculate the pH of pure water:

$$pH = -\log [H_3O^+]$$
$$pH = -\log [1.0 \times 10^{-7}]$$
$$pH = -[-7]$$
$$pH = 7$$

The pH of pure water is 7. Chemists call this point on the pH scale *neutral.* A solution is called *acidic* if it has a larger $[H_3O^+]$ than water and a smaller pH value than 7. A *basic* solution has a smaller $[H_3O^+]$ than water and a larger pH value than 7.

The pH scale really has no end. You can have a solution of pH that registers less than 0. (A 10 M HCl solution, for example, has a pH of −1.) However, the 0 to 14 range is a convenient range to use for weak acids and bases and for dilute solutions of strong acids and bases. Figure 11-3 shows the pH scale.

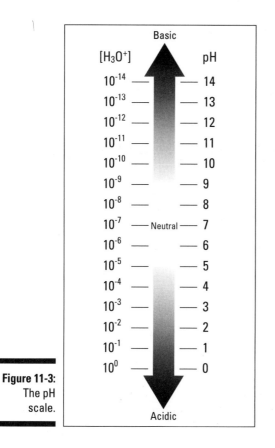

Figure 11-3:
The pH scale.

The $[H_3O^+]$ of a 2.0 M acetic acid solution is 6.0×10^{-3}. Looking at the pH scale, you see that this solution is acidic. Now calculate the pH of this solution:

$pH = -\log [H_3O^+]$

$pH = -\log [6.0 \times 10^{-3}]$

$pH = -[-2.22]$

$pH = 2.22$

In the section "Playing both parts: Amphoteric water," I explain that the K_w expression enables you to calculate the $[H_3O^+]$ if you have the $[OH^-]$. Another equation, called the *pOH,* can be useful in calculating the pH of a solution. The pOH is the negative logarithm of the $[OH^-]$. You can calculate the pOH of a solution just like the pH by taking the negative log of the hydroxide ion concentration. If you use the K_w expression and take the negative log of both sides, you get 14 = pH + pOH. This equation makes going from pOH to pH quite easy.

Just as you can convert from $[H_3O^+]$ to pH, you can also go from pH to $[H_3O^+]$. To do this, you use what's called the *antilog relationship,* which is

$$[H_3O^+] = 10^{-pH}$$

Human blood, for example, has a pH of 7.3. Here's how you calculate the $[H_3O^+]$ from the pH of blood:

$$[H_3O^+] = 10^{-pH}$$
$$[H_3O^+] = 10^{-7.3}$$
$$[H_3O^+] = 5.01 \times 10^{-8}$$

The same procedure can be used to calculate the $[OH^-]$ from the pOH.

Antacids: Good, basic chemistry

Go to any drugstore or grocery store and look at the shelves upon shelves of antacids. They represent acid-base chemistry in action!

The stomach secretes hydrochloric acid in order to activate certain enzymes (biological catalysts) in the digestion process. But sometimes the stomach produces too much acid, or the acid makes its way up into the esophagus (leading to heartburn or acid reflux), and you need to neutralize the excess acid with — you guessed it — a base. The basic formulations that are sold to neutralize this acid are called *antacids.* Antacids include the following compounds as active ingredients:

✔ **Bicarbonates** — $NaHCO_3$ and $KHCO_3$

✔ **Carbonates** — $CaCO_3$ and $MgCO_3$

✔ **Hydroxides** — $Al(OH)_3$ and $Mg(OH)_2$

Trying to select the "best" antacid for occasional use can be complicated. Certainly price is a factor, but the chemical nature of the bases can also be a factor. For example, individuals with high blood pressure may want to avoid antacids containing sodium bicarbonate because the sodium ion tends to increase blood pressure. Individuals concerned about loss of calcium from the bones, or *osteoporosis,* may want to use an antacid containing calcium carbonate. However, both calcium carbonate and aluminum hydroxide can cause constipation if used in large doses. On the other hand, large doses of both magnesium carbonate and magnesium hydroxide can act as laxatives. Selecting an antacid can really be a balancing act!

Substances commonly found in everyday life cover a wide range of pH values. Table 11-4 lists some common substances and their pH values.

Table 11-4	Average pH Values of Some Common Substances
Substance	*pH*
Oven cleaner	13.8
Hair remover	12.8
Household ammonia	11.0
Milk of magnesia	10.5
Chlorine bleach	9.5
Seawater	8.0
Human blood	7.3
Pure water	7.0
Milk	6.5
Black coffee	5.5
Soft drinks	3.5
Aspirin	2.9
Vinegar	2.8
Lemon juice	2.3
Auto battery acid	0.8

To sustain life, human blood must stay within about +/–0.2 pH units of 7.3, a narrow range. Many things in our environment, such as foods and hyperventilation, can act to change the pH of our blood. Buffers help to regulate blood pH and keep it in the 7.1 to 7.5 range.

Controlling pH with Buffers

Buffers, or *buffer solutions* as they're sometimes called, resist a change in pH caused by the addition of acids or bases. Obviously, the buffer solution must contain something that reacts with an acid — a base. Something else in the buffer solution reacts with a base — an acid. In general, buffers come in two types:

 ✔ **Mixtures of weak acids and bases:** The mixtures of weak acids and bases may be conjugate acid-base pairs (such as H_2CO_3/HCO_3^-) or nonconjugate acid-base pairs (such as NH_4^+/CH_3COO^-). (For more info about conjugate acid-base pairs, see "Competing for protons: Bronsted-Lowry acid-base reactions," earlier in this chapter.)

In the body, conjugate acid-base pairs are more common. In the blood, for example, the carbonic acid/bicarbonate pair helps to control the pH. This buffer can be overcome, though, and some potentially dangerous situations can arise. If a person exercises strenuously, lactic acid from the muscles is released into the bloodstream. If there's not enough bicarbonate ion to neutralize the lactic acid, the blood pH drops, and the person is said to be in *acidosis*. Diabetes may also cause acidosis. On the other hand, if a person hyperventilates (breathes too fast), she breathes out too much carbon dioxide. The carbonic acid level in the blood is reduced, causing the blood to become too basic. This condition, called *alkalosis,* can be very serious.

✔ **Amphoteric species:** Amphoteric species may also act as buffers by reacting with an acid or a base. (For an example of an amphoteric species, see "Playing both parts: Amphoteric water," earlier in this chapter.) The bicarbonate ion (HCO_3^-) and the monohydrogen phosphate ion (HPO_4^{-2}) are amphoteric species that neutralize both acids and bases. Both of these ions are also important in controlling the blood's pH.

Acids with bad press: An introduction to acid rain

Over the past decade or so, acid rain has emerged as a great environmental problem. Natural rainwater is somewhat acidic (around pH 5.6) due to the absorption of carbon dioxide from the atmosphere and the creation of carbonic acid. However, when acid rain is mentioned in the press, it usually refers to rain in the pH 3 to 3.5 range.

The two major causes of acid rain are automotive and industrial pollution. In the automobile's internal combustion engine, nitrogen in the air is oxidized to various oxides of nitrogen. These nitrogen oxides, when released into the atmosphere, react with water vapor to form nitric acid (HNO_3).

In fossil fuel power plants, oxides of sulfur are formed from the burning of the sulfur impurities commonly found in coal and petroleum. These oxides of sulfur, if released into the atmosphere, combine with water vapor to form both sulfuric and sulfurous acids (H_2SO_4 and H_2SO_3). Oxides of nitrogen are also produced in these power plants.

These acids fall to earth in the rain and cause a multitude of problems. They dissolve the calcium carbonate of marble statues and monuments. They decrease the pH of lake water to such a degree that fish can no longer live in the lakes. They cause whole forests to die or become stunted. They react with the metals in cars and buildings.

Industrial controls have been somewhat effective in reducing the problem, but acid rain is still a major environmental issue. (See Chapter 18 for more info about acid rain.)

Part III

Blessed Be the Bonds That Tie

The 5th Wave   By Rich Tennant

PROF. BLOWFISH HAD THE REPUTATION OF BEING SOMEWHAT UNAPPROACHABLE.

In this part . . .

Chemists operate in the microscopic world of atoms and molecules. I show you in this part the modern atomic theory — the quantum theory and how a radio wave and a locomotive are similar. I then give you the opportunity to explore chemical bonding — the ionic bonds that hold together the ions in a salt crystal and the covalent bonds that hold together a sugar molecule.

This part also shows you how to predict the shape of molecules. In chemistry, especially biochemistry, shape really does matter, and you discover how to determine the shape of molecules. You also uncover hybridization and molecular orbitals, theories about bonding. I revisit the periodic table and discuss periodic trends and finally end up talking about intermolecular forces, those forces that allow water to help support life on earth and make your hair curly or straight.

Chapter 12

Where Did I Put That Electron? Quantum Theory

The development of Rutherford's nuclear model that I discuss in Chapter 4 was an important step toward understanding the atom. In this model, protons and neutrons are located in a dense central core of the atom called the *nucleus,* while the electrons are located outside the nucleus. These electrons interact (or react) with electrons from other atoms. The nuclei are really only indirectly involved in chemical reactions.

Because the behavior of electrons is so very important, understanding this behavior is crucial. Scientists discovered that the behavior of electrons is very closely related to the behavior of light. The investigation of light and electrons led to a fundamental revision of physics known as quantum mechanics. This chapter gives you just the basics you may need to know about quantum mechanics during your Chem I class. As you venture deeper into your chemistry studies, you'll encounter more in-depth and complicated matters about quantum mechanics.

Facing the Concepts of Matter and Light

Before you can understand the behavior of the electron, you need to look first at the properties of light. Here I define a few terms and relationships concerning light, and then I focus on spectroscopy as the interaction of light and electrons.

Understanding the components

To begin to grasp the nature of electrons, examining the nature of light is necessary. Visible light, X-rays, microwaves, radio waves, and so on are all various forms of electromagnetic radiation. *Electromagnetic radiation,* sometimes referred to as *radiant energy,* carries energy through space. If that space is a vacuum, all types of electromagnetic radiation travel at the speed of light, denoted by *c,* 3.00×10^8 m/s. These waves of electromagnetic radiation also have the following three properties (see Figure 12-1):

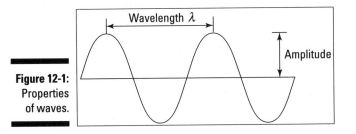

✔ **Amplitude:** The *amplitude* of a wave is the height of the wave from the midpoint or baseline of the wave to its peak (highpoint). Think of it as a measure of the strength of the electromagnetic radiation. For the visible part of the spectrum, you interpret the amplitude as the intensity or brightness of the light.

✔ **Wavelength:** *Wavelength,* λ (lambda), is the distance between two identical adjacent points of a wave, such as peaks (high points) or troughs (low points). You may express the wavelength in any unit of length, although many times chemists choose a specific unit for a specific type of electromagnetic radiation (meters for radio and TV waves or angstroms [10^{-10} m] for X-rays).

✔ **Frequency:** The *frequency,* ν (nu), is the number of waves that pass a given point during a specified time interval. Frequency has units of cycles per time, but chemists generally accept the cycles part as understood and omit it, leaving frequency expressed as reciprocal time (1/time). The SI base unit for time is the second, so frequency is measured in cycles per second and has the units of 1/s or s^{-1}.

Because these electromagnetic waves are all traveling at the speed of light, the wavelength and the frequency of the light have a direct relationship. If the wavelength is short, many waves pass a reference point per given amount of time and the frequency is high. If the wavelength is long, fewer waves pass the reference point during the same amount of time and the frequency is lower. You can see the relationship between the wavelength and frequency with this equation:

$$\lambda \nu = c$$

where λ is the wavelength (in meters), v is the frequency (in s^{-1}), and c is the speed of light (in m/s).

TIP

In working problems with this direct relationship between wavelength and frequency, expressing the wavelength, frequency, and speed of light in the units of meters, cycles per second, and meters per second is easiest.

Figure 12-2 shows the electromagnetic spectrum in order of increasing wavelength. Notice that gamma rays have the shortest wavelength (highest frequency) and radio waves have the longest wavelength (lowest frequency). Also notice the visible spectrum (the part of the electromagnetic spectrum that you can detect with your eyes) is a relatively small part of the entire electromagnetic spectrum. Scientists generally express the wavelength of the visible part of the spectrum in nanometers (10^{-9} meters), and the wavelength ranges from violet at about 400 nm to red at 750 nm.

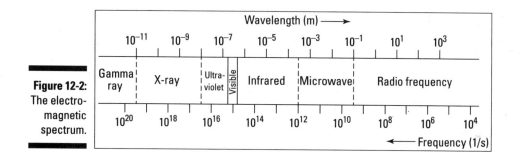

Figure 12-2:
The electro-
magnetic
spectrum.

As mentioned previously, frequency is usually in cycles per second (or simply s^{-1}). This unit of cycles per second is a hertz (Hz), and thus frequency can be in hertz or as any combination with SI prefixes (such as kilohertz). Chemists commonly express the frequency of radio stations in this fashion. Thus, 900 on the AM dial would be 900 kHz (kilohertz) or 900,000 s^{-1}.

Spectroscopy

In the early 1900s, Max Planck, a German scientist, found that energy came in discrete units. Planck labeled each of these units a *quantum*. He was able to relate the energy of these quanta to the frequency of the light with a simple constant. His theory of quantized energy of the electrons helped explain the fact that elements sometimes gave off light of certain distinct wavelengths. This relationship is

$$E = hv$$

E is the energy of the radiation in joules, h is Planck's constant (6.626×10^{-34} J $\times$ s), and v is the frequency of the radiation in s^{-1}.

The energy of a photon may also relate to its wavelength if you combine the following equations:

$$\lambda v = c \text{ and } E = h v$$

to get

$$E = \frac{hc}{\lambda}$$

Einstein extended Planck's work to say that particles of light known as *photons* carry Planck's quanta of energy. Thus, light has both wave properties (λ and v) and particle (photon) properties. This dual nature of light is very important in helping you understand the behavior of light, and a better understanding of the behavior of light leads to a better understanding of how light interacts with electrons.

According to Planck's theory, matter could only release or absorb energy in multiples of this quantum (packet) of energy. It's like a one-legged person going up the stairs. The individual can rest on one stair or another but not between the stair steps. Planck's theory means that the energy is *quantized;* that is, only certain values (multiples of the quantum) are allowed. A logical extension of Planck's theory is that the electrons in an atom can have only certain quantized energies associated with them.

Support for Planck's theory is found in the emission and absorption of elements.

Emission spectra

Light from most sources contains many different frequencies. You can separate these different frequencies into a *spectrum* (a range or band of wavelengths) using a prism or diffraction grating. The type of spectrum that you get depends upon the source of the light. For example, light from the sun or some other very hot object produces a *continuous spectrum,* a spectrum consisting of all frequencies of light.

When you excite elements in the gas phase by heating them or using electricity, the elements also emit light at various frequencies. The separation of this light yields a spectrum with only certain distinct frequencies, not a continuous spectrum. These separate frequencies appear as separate lines in the spectrum, so this type of spectrum is named a *line spectrum.* Because the spectrum results from individual atoms emitting light, it's also called an *emission spectrum.* The lines observed are characteristic of the element; each element gives a different combination of lines.

Absorption spectra

Another type of spectrum is an *absorption spectrum*. To yield this type of spectrum, the atoms must be in the gaseous state (as for an emission spectrum). However, instead of the thermal or electrical energy needed to produce an emission spectrum, light from a continuous spectrum source is used. The passage of this light through the gaseous atoms results in the absorption of some of the frequencies. The absorption spectrum results from this transmitted light.

If you use atoms of the same element for an emission spectrum and an absorption spectrum, you find that the lines in the emission spectrum correspond to the absorbed frequencies in the absorption spectrum. This correspondence of frequencies indicates that a relationship must exist between the emission of light and the absorption of light. Because specific frequencies are involved, according to Planck's theory, the element can emit or absorb only specific quanta of energy.

The first step toward understanding the spectra of the elements came from an examination of the spectrum of the simplest element — hydrogen. The spectrum of hydrogen consists of groups of lines in different regions of the spectrum. Analysis of this repeated pattern led Niels Bohr to develop a theory to explain not only the spectrum of the hydrogen atom but also a structure of the atom itself (see the next section for more information).

Grasping Bohr's Atomic Model

Niels Bohr, attempting to explain the line spectrum of hydrogen, developed a model combining the concepts developed by Planck and Einstein (addressed previously under "Spectroscopy"). Bohr assumed that the atom contained a nucleus and that the electrons circled the nucleus in circular orbits. He had to introduce three postulates in order to make his model consistent with observations:

- ✔ The electron in a hydrogen atom may only occupy orbits of certain radii that corresponded to certain discrete energies.

- ✔ While an electron is in an allowed energy state (orbit), it does not radiate energy and it remains in that orbit without spiraling into the nucleus.

- ✔ An electron may move from one energy state to another by absorbing or releasing energy. The energy needed is the difference between one energy level and another and is equal to a photon, $E = h\nu$.

In Bohr's atomic model, the number n designates the energy state (level) occupied by an electron. The smaller the n value, the smaller the radius (closer to the nucleus) and the lower the energy associated with the electron. An electron that occupies its lowest possible energy level is in its *ground state*. It may absorb energy to give an absorption spectra, and when it does, the energy absorbed must equal the difference between the ground state energy and a higher energy level. The electron will then move from the lower state to the higher one, which is an *excited state*. The electron can return to its ground state by emitting a photon of energy equal to the energy difference in these two states. These transitions explain the emission line spectra that occur when atoms emit distinct lines of frequencies (which I discuss in "Emission spectra").

The Bohr model works well in predicting the spectral lines of hydrogen but not nearly as well for any other atom. In other atoms, the additional electron-nucleus attractions and the electron-electron repulsions introduce complications that Bohr's model can't handle. In addition, the critical assumption of the Bohr model, that electrons travel in discrete orbits, was found to be incorrect. The motion of the electrons is far more complex. It was up to other scientists to build upon this notion. The following sections identify a couple other scientists' work with the atomic model.

De Broglie's contribution

The work of Planck and Einstein firmly established the dual nature of radiant energy. Light had properties of both particles (photons) and waves. Louis de Broglie in the early 1920s proposed the idea that if light waves had properties of particles, then particles could have the properties of waves (wavelength and so on). De Broglie put forth the idea that a simple equation relates the wavelength, λ, associated with an object, the object's mass, m, and its velocity, v:

$$\lambda = \frac{h}{mv}$$

where h is Planck's constant. This equation is the de Broglie relationship.

Don't confuse ν (frequency) with v (velocity).

De Broglie introduced the concept that all matter, even you, has both wave-like and particle-like properties. Looking at the mathematical relationship, note that the wavelength is inversely proportional to the mass. The larger the mass, the smaller the wave nature becomes and the more important the particle nature of the object. Therefore, you can describe a train, a car, or even a gold ball in terms of its wave nature, but the description wouldn't be very useful, because the wave nature of these objects is very small. They can best be described in terms of their particle nature. However, an electron, with its extremely tiny mass, has a much larger wave nature associated with it and may be better described with its wave properties.

De Broglie's hypothesis of the dual properties of electrons gained acceptance in 1927 when scientists found that a stream of electrons passing through a crystal exhibited the same behavior as a stream of X-rays. The scientific community began to embrace the concept of the dual properties of both energy and matter.

Heisenberg's contribution

Planck's theory, even the wave portion, was still based on classical physics, but along came Heisenberg and his introduction of uncertainty.

The concept of the wave-like nature of the electron had a major impact on modern physics and the view of the world. In classical mechanics, scientists treated the electron as a particle with a certain measurable position and momentum. If they knew the position and momentum, physicists could then calculate the position and momentum at points in the past and in the future with a great deal of accuracy. However, the German physicist Werner Heisenberg realized that with something as small as an electron, the location of the particle can't be measured with accuracy.

Heisenberg formalized his ideas in his 1927 paper on the uncertainty principle, which states that accurately determining both the position and momentum of a particle simultaneously is impossible. You may determine one, but in doing so your measurement techniques influence and change the other. Heisenberg's uncertainty principle means that you can't determine the precise position of the electron; the best you can do is assign probabilities to volumes of space in which you may find electrons.

Understanding the Quantum Mechanical Model

In 1926, Erwin Schrödinger introduced a mathematical relationship called a *wave equation* that takes into account both the wave and particle nature of the electron. That equation was instrumental in the development of a completely new way of representing the behavior of subatomic particles. This new field is *wave mechanics* or *quantum mechanics*.

Solving Schrödinger's wave equation for the electron in hydrogen results in the generation of a mathematical expression called a *wave function,* represented by Greek ψ (psi). The square of this wave function, ψ^2, provides you with information concerning the probability of finding an electron at a certain location in hydrogen. Because of the Heisenberg uncertainty principle, Schrödinger represented this information in terms of a volume of space in which it is more probable to find the electron (not in terms of distinct orbits as described by Bohr).

In order to solve Schrödinger's wave equation for a specific electron in an atom, you must specify three numbers, called *quantum numbers.* In addition, you must use a fourth quantum number to differentiate between the two electrons sharing a specific volume of space. The four quantum numbers and their mathematically allowed values are as follows:

- **Principal quantum number:** This number, n, denotes the energy level or shell that the electron occupies. The values of n are positive integers 1, 2, 3, 4, and so on. The larger the value of n, the greater the average distance the electron is from the nucleus and the higher the energy associated with the electrons.

- **Angular momentum quantum number:** This number, l, designates the shape of the volume of space within an energy level that most likely contains the electrons. This volume of space is a *subshell.* The allowed values of l range from 0 to $n - 1$ in integer steps. For example, if $n = 1$, then the only allowed value of l is 0. If $n = 3$, then l may have the values of 0, 1, and 2 ($= 3 - 1$). The various values of l correspond to different shapes. In addition, each subshell contains one or more orbitals, volumes of space in which electrons may be found. The orbitals contained in a specific subshell have the same energy. An individual orbital may contain a maximum of two electrons.

- **Magnetic quantum number:** This number, m_l, describes the orientation of the orbital in space. It may have values of $-l$ to 0 to $+l$ in integer steps. For example if $l = 0$, then the only allowed value for m_l is 0. If $l = 3$, then m_l may have values of $-3, -2, -1, 0, +1, +2$, and $+3$. This means that when considering a subshell of $l = 3$, seven different orbitals (seven different allowed values of m_l) are within that subshell.

- **Electron spin quantum number:** This number, m_s, designates the direction of the magnetic field that the electron is generating. It may align either with or against the field of the atom. Therefore, two values are allowed, $+\frac{1}{2}$ and $-\frac{1}{2}$.

Probabilities: In science and the real world

The idea of probabilities associated with the location of an electron may seem strange, but you deal with probabilities of location in your macroscopic life every day. Do you ever ride trains? Can you predict the *exact* time the train will arrive at the station? Probably not. However, you can easily establish an interval of time (perhaps ten minutes or so) that is most probable for the train to arrive. Your friends may know that the movie you plan to attend is showing from 7:00 to 9:15 p.m. Does this mean that they know without a doubt that you'll be sitting in that movie theater at 7:05 that evening? Probably not. You may be in the restroom or standing in line at the concession stand. It is most probable that you are in that movie theater, but not certain. Scientists use probabilities in a similar way in the microscopic world.

Under the quantum mechanical model, chemists and physicists speak of probability densities or electron densities instead of an exact location. Scientists commonly calculate probability densities at a 90 percent probability level, meaning that the size of the three-dimensional figure is such that you can probably find the electron within that volume of space 90 percent of the time. Scientists sometimes call these electron density plots *electron clouds* and quite commonly use them in representing the probabilities of finding various electrons within an atom.

Chapter 13

Opposites Do Attract: Ionic Bonding

*I*f I had to point to the one thing that made me want to major in chemistry, it would be the reactions of salts. I remember the day clearly: It was the second half of general chemistry, and I was doing qualitative analysis (finding out what's in a sample) of salts. I really enjoyed the colors of the compounds formed in the reactions I was doing, and the labs were fun and challenging. I was hooked.

In this chapter, I introduce you to *ionic bonding,* the type of bonding that holds salts together. I discuss simple ions and polyatomic ions: how they form and how they combine. I also show you how to predict the formulas of ionic compounds and how chemists detect ionic bonds. You may not decide to devote your life to chemistry after reading this chapter, but it will definitely help you get through your class.

Magically Bonding Ions: Sodium + Chlorine = Table Salt

Sodium is a fairly typical metal. It's silvery, soft, and a good conductor. It's also highly reactive. Sodium is normally stored under oil to keep it from reacting with the water in the atmosphere. If you melt a freshly cut piece of sodium and put it into a beaker filled with greenish-yellow chlorine gas, something very impressive happens. The molten sodium begins to glow with a white light that gets brighter and brighter. The chlorine gas swirls, and soon the color of the gas begins to disappear. In a couple of minutes, the reaction is over, and the beaker can be safely uncovered. You find a white crystalline substance, table salt (NaCl), deposited on the inside of the beaker.

In the following sections, I show you what happens during the chemical reaction to create table salt and, more importantly, why it occurs. Understanding these concepts will go a long way in your investigation into ionic bonding.

Meeting the components

If you really stop and think about it, the process of creating table salt is pretty remarkable. You take the following two substances that are both very hazardous (the Germans used chlorine gas against the opposing troops during World War I), and from them you make a substance that's necessary for life.

- ✔ **Sodium** is an alkali metal, a member of the IA family on the periodic table. The Roman numerals at the top of the A families show the number of valence electrons (s and p electrons in the outermost energy level) in the particular element (see Chapter 5 for details). So sodium has 1 valence electron and 11 total electrons because its atomic number is 11.

  You can use an energy-level diagram to represent the distribution of electrons in an atom. Sodium's energy-level diagram is shown in Figure 13-1. (If energy-level diagrams are new to you, check out Chapter 4. A number of minor variations are commonly used in writing energy-level diagrams, so don't worry if the diagrams in Chapter 4 are slightly different than the ones I show you here.)

  Chlorine is a member of the halogen family — the VIIA family on the periodic table. It has 7 valence electrons and a total of 17 electrons. The energy-level diagram for chlorine is also shown in Figure 13-1.

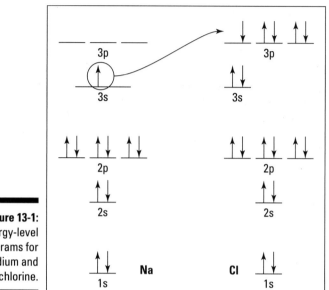

Figure 13-1:
Energy-level
diagrams for
sodium and
chlorine.

If you want, instead of using the bulky energy-level diagram to represent the distribution of electrons in an atom, you can use the electron configuration. (For a complete discussion of electron configurations, see Chapter 4.) Write, *in order,* the energy levels being used, the orbital types (s, p, d, and so on), and — in superscript — the number of electrons in each orbital. Here are the electronic configurations for sodium and chlorine:

Sodium (Na) $1s^2 2s^2 2p^6 3s^1$

Chlorine (Cl) $1s^2 2s^2 2p^6 3s^2 3p^5$

Understanding the reaction

The noble gases are the VIIIA elements on the periodic table. They're extremely unreactive because each atom's *valence* energy level (outermost energy level) is filled. Achieving a filled (complete) valence energy level is a driving force in nature in terms of chemical reactions, because that's when elements become stable, or *unreactive.* They don't lose, gain, or share electrons.

The other elements in the A families on the periodic table do gain, lose, or share valence electrons in order to fill their valence energy level and become stable. Because this process, in most cases, involves filling the outermost s and p orbitals, it's sometimes called the *octet rule* — elements gain, lose, or share electrons to reach a full octet (eight valence electrons: two in the s orbital and six in the p orbital).

Sodium's role

Sodium has one valence electron; by the octet rule, it becomes stable when it has eight valence electrons. Two possibilities exist for sodium to become stable: It can gain seven more electrons to fill energy level 3, or it can lose the one 3s electron so that energy level 2 (which is already filled with eight electrons) becomes the valence energy level. In general, the loss or gain of one, two, or sometimes even three electrons can occur, but an element doesn't ordinarily lose or gain more than three electrons. So to gain stability, sodium loses its 3s electron. At this point, it has 11 protons (11 positive charges) and 10 electrons (10 negative charges). The once-neutral sodium atom now has a single positive charge [11(+) plus 10(−) equals 1+]. It's now an *ion,* an atom that has a charge due to the loss or gain of electrons. And ions that have a positive charge (such as sodium) due to the loss of electrons are called *cations.* You can write an electron configuration for the sodium cation:

$$Na^+ \quad 1s^2 2s^2 2p^6$$

The sodium ion (cation) has the same electron configuration as neon, so it's *isoelectronic* with neon. So has sodium become neon by losing an electron? No. Sodium still has 11 protons, and the number of protons determines the identity of the element.

The neutral sodium atom and the sodium cation have the difference of one electron. In addition, their chemical reactivities are different *and* their sizes are different. The cation is smaller. The outermost energy level determines the size of an atom or ion (or, in this case, cation). Because sodium loses an entire energy level to change from an atom to a cation, the cation is smaller.

Chlorine's role

Chlorine has seven valence electrons. To obtain its full octet, it must lose the seven electrons in energy level 3 or gain one at that level. Because elements don't generally gain or lose more than three electrons, chlorine must gain a single electron to fill energy level 3. At this point, chlorine has 17 protons (17 positive charges) and 18 electrons (18 negative charges). So chlorine becomes an ion with a single negative charge (Cl^-). The neutral chlorine atom becomes the chloride ion. Ions with a negative charge due to the gain of electrons are called *anions.* The electronic configuration for the chloride anion is

$$Cl^- \quad 1s^2 2s^2 2p^6 3s^2 3p^6$$

The chloride anion is isoelectronic with argon. The chloride anion is also slightly larger than the neutral chlorine atom. To complete the octet, the one electron gained went into energy level 3, but now 17 protons are attracting 18 electrons. The attractive force on each electron has been reduced slightly, and the electrons are free to move outward a little, making the anion a little larger.

In general, a cation is smaller than its corresponding atom, and an anion is slightly larger.

Ending up with a bond

Sodium can achieve its full octet and stability by losing an electron. Chlorine can fill its octet by gaining an electron. If the two are in the same container, then the electron sodium loses can be the same electron chlorine gains. I show this process in Figure 13-1, indicating that the 3s electron in sodium is transferred to the 3p orbital of chlorine.

The transfer of an electron creates ions — cations (positive charge) and anions (negative charge) — and opposite charges attract each other. The Na^+ cation attracts the Cl^- anion and forms the compound NaCl, or table salt. This is an example of an *ionic bond,* which is a *chemical bond* (a strong attractive force that keeps two chemical elements together) that comes from the *electrostatic attraction* (attraction of opposite charges) between cations and anions.

The compounds that have ionic bonds are commonly called *salts.* In sodium chloride, a crystal is formed in which each sodium cation is surrounded by six different chloride anions, and each chloride anion is surrounded by six different sodium cations. The crystal structure is shown in Figure 13-2.

Figure 13-2:
Crystal
structure
of sodium
chloride.

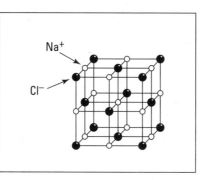

Notice the regular, repeating structure. Different types of salts have different crystal structures. Cations and anions can have more than one unit of positive or negative charge if they lose or gain more than one electron. In this fashion, many different kinds of salts are possible.

Ionic bonding, the bonding that holds the cations and anions together in a salt, is one of the two major types of bonding in chemistry. The other type, *covalent bonding,* is described in Chapter 14. Grasping the concepts involved in ionic bonding makes understanding covalent bonding much easier.

Identifying Positive and Negative Ions: Cations and Anions

The basic process that occurs when sodium chloride is formed also occurs when other salts are formed. A metal loses electrons, and a nonmetal gains those electrons. Cations and anions are formed, and the electrostatic attraction between the positives and negatives brings the particles together and creates the ionic compound.

A metal reacts with a nonmetal to form an ionic bond.

You can often determine the charge an ion normally has by the element's position on the periodic table. For example, all the alkali metals (the IA elements) lose a single electron to form a cation with a 1+ charge. In the same way, the alkaline earth metals (IIA elements) lose two electrons to form a 2+ cation. Aluminum, a member of the IIIA family, loses three electrons to form a 3+ cation.

By the same reasoning, the halogens (VIIA elements) all have seven valence electrons. All the halogens gain a single electron to fill their valence energy level. And all of them form an anion with a single negative charge. The VIA elements gain two electrons to form anions with a 2– charge, and the VA elements gain three electrons to form anions with a 3– charge.

Table 13-1 shows the family, element, ion name, and ion symbol for some common monoatomic (one atom) cations, and Table 13-2 gives the same information for some common monoatomic anions.

Table 13-1   Some Common Monoatomic Cations

Family	Element	Ion Name	Ion Symbol
IA	Lithium	Lithium cation	Li^+
	Sodium	Sodium cation	Na^+
	Potassium	Potassium cation	K^+
IIA	Beryllium	Beryllium cation	Be^{2+}
	Magnesium	Magnesium cation	Mg^{2+}
	Calcium	Calcium cation	Ca^{2+}
	Strontium	Strontium cation	Sr^{2+}
	Barium	Barium cation	Ba^{2+}
IB	Silver	Silver cation	Ag^+
IIB	Zinc	Zinc cation	Zn^{2+}
IIIA	Aluminum	Aluminum cation	Al^{3+}

Table 13-2   Some Common Monoatomic Anions

Family	Element	Ion Name	Ion Symbol
VA	Nitrogen	Nitride anion	N^{3-}
	Phosphorus	Phosphide anion	P^{3-}
VIA	Oxygen	Oxide anion	O^{2-}
	Sulfur	Sulfide anion	S^{2-}
VIIA	Fluorine	Fluoride anion	F^-
	Chlorine	Chloride anion	Cl^-
	Bromine	Bromide anion	Br^-
	Iodine	Iodide anion	I^-

Determining the number of electrons that members of the transition metals (the B families) lose is more difficult. In fact, many of these elements lose a varying number of electrons so that they form two or more cations with different charges.

The electrical charge that an atom achieves is sometimes called its *oxidation state*. Many of the transition metal ions have varying oxidation states. Table 13-3 shows some common transition metals that have more than one oxidation state.

Table 13-3		Some Common Metals with More Than One Oxidation State	
Family	*Element*	*Ion Name*	*Ion Symbol*
VIB	Chromium	Chromium(II) or chromous	Cr^{2+}
		Chromium(III) or chromic	Cr^{3+}
VIIB	Manganese	Manganese(II) or manganous	Mn^{2+}
		Manganese(III) or manganic	Mn^{3+}
VIIIB	Iron	Iron(II) or ferrous	Fe^{2+}
		Iron(III) or ferric	Fe^{3+}
	Cobalt	Cobalt(II) or cobaltous	Co^{2+}
		Cobalt(III) or cobaltic	Co^{3+}
IB	Copper	Copper(I) or cuprous	Cu^{+}
		Copper(II) or cupric	Cu^{2+}
IIB	Mercury	Mercury(I) or mercurous	Hg_2^{2+}
		Mercury(II) or mercuric	Hg^{2+}
IVA	Tin	Tin(II) or stannous	Sn^{2+}
		Tin(IV) or stannic	Sn^{4+}
	Lead	Lead(II) or plumbous	Pb^{2+}
		Lead(IV) or plumbic	Pb^{4+}

Notice that these cations can have more than one name. The current way of naming ions is to use the metal name, such as chromium, followed in parentheses by the ionic charge written as a Roman numeral, such as (II). An older way of naming ions uses *-ous* and *-ic* endings. When an element has more than one ion — chromium, for example — the ion with the lower oxidation state (lower numerical charge, ignoring the + or –) is given an *-ous* ending, and the ion with the higher oxidation state (higher numerical charge) is given an *-ic* ending. So for chromium, the Cr^{2+} ion is named *chromous* and the Cr^{3+} ion is named *chromic*. (See the section "Naming Ionic Compounds," later in this chapter, for more on naming ions.)

Grasping Polyatomic Ions

Ions aren't always *monoatomic,* composed of just one atom. Ions can also be *polyatomic,* composed of a group of atoms. For example, refer to Table 13-3. Notice anything about the mercury(I) ion? Its ion symbol, Hg_2^{2+}, shows that two mercury atoms are bonded together. This group has a 2+ charge, with each mercury cation having a 1+ charge. The mercurous ion is classified as a polyatomic ion.

Polyatomic ions are treated the same as monoatomic ions (see "Naming Ionic Compounds," later in this chapter). Table 13-4 lists some important polyatomic ions. Many of the compounds you'll encounter in chemistry contain polyatomic ions.

Table 13-4	Some Important Polyatomic Ions
Ion Name	*Ion Symbol*
Sulfate	SO_4^{2-}
Sulfite	SO_3^{2-}
Nitrate	NO_3^-
Nitrite	NO_2^-
Hypochlorite	ClO^-
Chlorite	ClO_2^-
Chlorate	ClO_3^-
Perchlorate	ClO_4^-
Acetate	$C_2H_3O_2^-$
Chromate	CrO_4^{2-}
Dichromate	$Cr_2O_7^{2-}$
Arsenate	AsO_4^{3-}
Hydrogen phosphate	HPO_4^{2-}
Dihydrogen phosphate	$H_2PO_4^-$
Bicarbonate or hydrogen carbonate	HCO_3^-
Bisulfate or hydrogen sulfate	HSO_4^-
Mercury(I)	Hg_2^{2+}

(continued)

Table 13-4 (continued)

Ion Name	Ion Symbol
Ammonium	NH_4^+
Phosphate	PO_4^{3-}
Carbonate	CO_3^{2-}
Permanganate	MnO_4^-
Cyanide	CN^-
Cyanate	OCN^-
Thiocyanate	SCN^-
Oxalate	$C_2O_4^{2-}$
Thiosulfate	$S_2O_3^{2-}$
Hydroxide	OH^-
Arsenite	AsO_3^{3-}
Peroxide	O_2^{2-}

The symbol for the sulfate ion, SO_4^{2-}, indicates that one sulfur atom and four oxygen atoms are bonded together and that the whole polyatomic ion has two extra electrons.

Putting Ions Together: Ionic Compounds

When an ionic compound is formed, the cation and anion attract each other, resulting in the formation of a salt (see "Magically Bonding Ions: Sodium + Chlorine = Table Salt," earlier in this chapter). An important thing to remember is that the compound must be *neutral* — it must have equal numbers of positive and negative charges. In the following sections I show you how to predict the formula of an ionic compound simply by considering the electronic configurations and/or the charges on the cations and anions.

Putting magnesium and bromine together

Suppose you want to know the *formula*, or composition, of the compound that results from reacting magnesium with bromine. You start by putting the two atoms side by side, with the metal on the left, and then adding their charges. Figure 13-3 shows this process. (Forget about the crisscrossing lines for now. Well, if you're really curious, check out the next section.)

Figure 13-3:
Figuring the
formula of
magnesium
bromide.

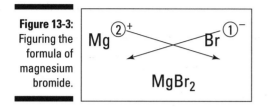

The electron configurations for magnesium and bromine are

Magnesium (Mg) $1s^22s^22p^63s^2$

Bromine (Br) $1s^22s^22p^63s^23p^64s^23d^{10}4p^5$

Magnesium, an alkaline earth metal, has two valence electrons that it loses to form a cation with a 2+ charge. The electron configuration for the magnesium cation is

Mg^{2+} $1s^22s^22p^6$

Bromine, a halogen, has seven valence electrons, so it gains one to complete its octet (eight valence electrons) and forms the bromide anion with a 1– charge. The electron configuration for the bromide anion is

Br^{1-} $1s^22s^22p^63s^23p^64s^23d^{10}4p^6$

Note that if the anion simply has 1 unit of charge, positive or negative, you normally don't write the 1; you just use the plus or minus symbol, with the 1 being understood. But for the example of the bromide ion, I use the 1.

The compound must be neutral; it must have the same number of positive and negative charges so that, overall, it has a zero charge. The magnesium ion has a 2+ charge, so it requires 2 bromide anions, each with a single negative charge, to balance the 2 positive charges of magnesium. So the formula of the compound that results from reacting magnesium with bromine is $MgBr_2$.

Applying the crisscross rule

A quick way to determine the formula of an ionic compound is to use the *crisscross rule*. The crisscross rule uses the ionic charges of the ions to predict the formula of the ionic compound. It doesn't work all the time, but it's a good way of checking your result using the previous method.

Refer to Figure 13-3 for an example of using this rule. Take the numerical value of the metal ion's superscript (forget about the charge symbol) and move it to the bottom right-hand side of the nonmetal's symbol — as a subscript. Then take the numerical value of the nonmetal's superscript and make it the subscript of the metal. (Note that if the numerical value is 1, it's just understood and not shown.) So in this example, you make magnesium's 2 a subscript of bromine and make bromine's 1 a subscript of magnesium (but because it's 1, you don't show it), and you get the formula $MgBr_2$.

So what happens if you react aluminum and oxygen? Figure 13-4 shows the crisscross rule used for this reaction.

Figure 13-4:
Figuring the
formula of
aluminum
oxide.

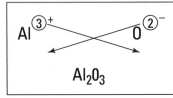

Compounds involving polyatomic ions work exactly the same way. For example, here's the compound made from the ammonium cation (NH_4^+) and the sulfide anion (S^{2-}):

$(NH_4)_2S$

Notice that because two ammonium ions (two positive charges) are needed to neutralize the two negative charges of the sulfide ion, the ammonium ion is enclosed in parentheses and a subscript 2 is added.

The crisscross rule works very well, but you have to be careful if both ions have the same numeral in the superscript. Suppose that you want to write the compound formed when calcium reacts with oxygen. Magnesium, an alkaline earth metal, forms a 2+ cation, and oxygen forms a 2– anion. So you might predict that the formula is

Mg_2O_2

But this formula is incorrect. After you use the crisscross rule, you need to reduce all the subscripts by a common factor, if possible. In this case, you divide each subscript by 2 and get the correct formula:

MgO

Naming Ionic Compounds

When you name inorganic compounds, you write the name of the metal first and then the nonmetal. Suppose, for example, that you want to name Li_2S, the compound that results from the reaction of lithium and sulfur. You first write the name of the metal, lithium, and then write the name of the nonmetal, adding an *-ide* ending so that *sulfur* becomes *sulfide*.

Li_2S lithium sulfide

Ionic compounds involving polyatomic ions follow the same basic rule: Write the name of the metal first, and then simply add the name of the nonmetal (with the polyatomic anions, adding the *-ide* ending is unnecessary).

$(NH_4)_2CO_3$ ammonium carbonate

K_3PO_4 potassium phosphate

When the metal involved is a transition metal with more than one oxidation state (see "Identifying Positive and Negative Ions: Cations and Anions," earlier in the chapter, for more info about that), the compound can be correctly named in more than one way. For example, suppose that you want to name the compound formed between the Fe^{3+} cation and the cyanide ion, CN^-. The preferred method is to use the metal name followed in parentheses by the ionic charge written as a Roman numeral: iron(III). But an older naming method that is still sometimes used (so knowing it is a good idea) uses *-ous* and *-ic* endings. The ion with the lower oxidation state (lower numerical charge, ignoring the + or –) gets an *-ous* ending, and the ion with the higher oxidation state (higher numerical charge) gets an *-ic* ending. So, because Fe^{3+} has a higher oxidation state than Fe^{2+}, it's called a *ferric ion*. So the compound can be named

$Fe(CN)_3$ iron(III) cyanide or ferric cyanide

Sometimes figuring out the charge on an ion can be a little challenging (and fun), so now I want to show you how to name $FeNH_4(SO_4)_2$.

I show you back in Table 13-4 that the sulfate ion has a 2– charge, and from the formula you can see that there are two of them. Therefore, you have a total of four negative charges. Table 13-4 also indicates that the ammonium ion has a 1+ charge, so you can figure out the charge on the iron cation.

Ion	*Charge*
Fe	?
NH_4	1+
$(SO_4)_2$	$(2-) \times 2$

Because you have a 4– for the sulfates and a 1+ for the ammonium, the iron must be a 3+ to make the compound neutral. So the iron is in the iron(III), or ferric, oxidation state. You can name the compound

$FeNH_4(SO_4)_2$ iron(III) ammonium sulfate or ferric ammonium sulfate

And, finally, if you have the name, you can derive the formula and the charge on the ions. For example, suppose that you're given the name *cuprous oxide*. You know that the cuprous ion is Cu^+ and the oxide ion is O^{2-}. Applying the crisscross rule, you get the following formula:

Cuprous oxide Cu_2O

Contrasting Electrolytes and Nonelectrolytes

When an ionic compound such as sodium chloride is put into water, the water molecules attract both the cations and anions in the crystal (the crystal is shown in Figure 13-2) and pull them into the solution. (In Chapter 14, I talk a lot about water molecules and show you why they attract the NaCl ions.) The cations and anions get distributed throughout the solution. Substances that conduct electricity in the molten state or when dissolved in water are called *electrolytes*. Substances that don't conduct electricity when in these states are called *nonelectrolytes*.

You can detect the presence of these ions by using an instrument called a conductivity tester. A *conductivity tester* tests whether water solutions of various substances conduct electricity. It's composed of a light bulb with two electrodes attached. The light bulb is plugged into a wall outlet, but it doesn't light until some type of conductor (substance capable of transmitting electricity) between the electrodes completes the circuit. (A finger also completes the circuit, so this experiment should be done carefully. If you're not careful, it can be a shocking experience!)

When you place the electrodes in pure water, nothing happens, because there's no conductor between the electrodes. Pure water is a nonconductor. But if you put the electrodes in the NaCl solution, the light bulb lights, because the ions conduct the electricity (carry the electrons) from one electrode to the other.

In fact, you don't even really need the water. If you were to melt pure NaCl (it requires a *lot* of heat!) and then place the electrodes into it, you'd find that the molten table salt also conducts electricity. In the molten state, the NaCl ions are free to move and carry electrons, just as they are in the saltwater solution.

Scientists can get some good clues as to the type of bonding in a compound by discovering whether a substance is an electrolyte or a nonelectrolyte. Ionically bonded substances act as electrolytes. But covalently bonded compounds (see Chapter 14), in which no ions are present, are commonly nonelectrolytes. Table sugar, or sucrose, is a good example of a nonelectrolyte. You can dissolve sugar in water or melt it, but it won't have conductivity. No ions are present to transfer the electrons. However, as you discover in Chapter 14, a few covalent compounds ionize (produce ions) when put into water.

Chapter 14

Sharing Nicely: Covalent Bonding

Sometimes when I'm cooking, I have one of my chemistry nerd moments and start reading the ingredients on food labels. I usually find lots of salts, such as sodium chloride, and lots of other compounds, such as potassium nitrate, that are all ionically bonded (see Chapter 13). But I also find many compounds, such as sugar, that aren't ionically bonded.

If no ions are holding a compound together, what does hold it together? What holds together sugar, vinegar, and even DNA? In this chapter, I discuss the other major type of bonding: covalent bonding. I explain the basics with an extremely simple covalent compound, hydrogen, and I tell you some cool stuff about one of the most unusual covalent compounds I know — water.

Eyeing Covalent Bond Basics

An *ionic bond* is a chemical bond that comes from the transfer of electrons from a metal to a nonmetal, resulting in the formation of oppositely charged ions — cations (positive charge) and anions (negative charge) — and the attraction between those oppositely charged ions. The driving force in this whole process is achieving a filled valence energy level, completing the atom's octet. (For a more complete explanation of this concept, see Chapter 13.)

But many other compounds exist in which electron transfer hasn't occurred. The driving force is still the same: achieving a filled valence energy level. But instead of achieving it by gaining or losing electrons, the atoms in these compounds *share* electrons. That's the basis of a *covalent bond.* The following sections provide more insight into covalent bonds.

Considering a hydrogen example

Hydrogen is #1 on the periodic table — upper left corner. The hydrogen found in nature is often not comprised of an individual atom. It's primarily found as H_2, a *diatomic* (two atom) element. (Taken one step further, because a *molecule* is a combination of two or more atoms, H_2 is called a *diatomic molecule.*)

Hydrogen has one valence electron. It'd love to gain another electron to fill its 1s energy level, which would make it *isoelectronic* with helium (because the two would have the same electronic configuration), the nearest noble gas. Energy level 1 can only hold two electrons in the 1s orbital, so gaining another electron fills it. The driving force of hydrogen is filling the valence energy level and achieving the same electron arrangement as the nearest noble gas.

Imagine one hydrogen atom transferring its single electron to another hydrogen atom. The hydrogen atom receiving the electron fills its valence shell and reaches stability while becoming an anion (H^-). However, the other hydrogen atom now has no electrons (H^+) and moves further away from stability. This process of electron loss and gain simply won't happen, because the driving force of *both* atoms is to fill their valence energy level. So the H_2 compound can't result from the loss or gain of electrons. What *can* happen is that the two atoms share their electrons. At the atomic level, this sharing is represented by the overlap of the electron orbitals (sometimes called *electron clouds*). The two electrons (one from each hydrogen atom) "belong" to both atoms. Each hydrogen atom feels the effect of the two electrons; each has, in a way, filled its valence energy level. A *covalent bond* is formed — a chemical bond that comes from the sharing of one or more electron pairs between two atoms. The overlapping of the electron orbitals and the sharing of an electron pair are represented in Figure 14-1a.

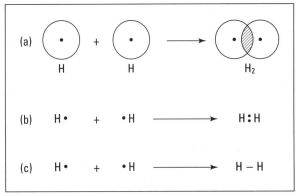

Figure 14-1:
The formation of a covalent bond in hydrogen.

Another way to represent this process is through the use of an *electron-dot formula*. In this type of formula, valence electrons are represented as dots surrounding the atomic symbol, and the shared electrons are shown between the two atoms involved in the covalent bond. The electron-dot formula representations of H_2 are shown in Figure 14-1b.

Most of the time, I use a slight modification of the electron-dot formula called the *Lewis structural formula;* it's basically the same as the electron-dot formula, but the shared pair of electrons (the covalent bond) is represented by a dash. The Lewis structural formula of diatomic hydrogen is shown in Figure 14-1c. (Check out the section, "Structural formula: Add the bonding pattern," for more about writing structural formulas of covalent compounds.)

In addition to hydrogen, six other elements are found in nature in the diatomic form: oxygen (O_2), nitrogen (N_2), fluorine (F_2), chlorine (Cl_2), bromine (Br_2), and iodine (I_2). So when I talk about oxygen gas or liquid bromine, I'm talking about the diatomic element (diatomic molecule).

For another example of using the electron-dot formula to represent the shared electron pair of a diatomic compound, look at bromine (Br_2), which is a member of the halogen family (see Figure 14-2). The two halogen atoms, each with seven valence electrons, share an electron pair and fill their octet.

Figure 14-2:
The covalent bond formation of Br_2.

$$:\overset{..}{\underset{..}{Br}}\cdot \quad + \quad :\overset{..}{\underset{..}{Br}}\cdot \quad \longrightarrow \quad :\overset{..}{\underset{..}{Br}}:\overset{..}{\underset{..}{Br}}:$$

$$\left(:\overset{..}{\underset{..}{Br}} - \overset{..}{\underset{..}{Br}}:\right)$$

Comparing covalent bonds with other bonds

Ionic bonding occurs between a metal and a nonmetal. Covalent bonding, on the other hand, occurs between two nonmetals. The properties of these two types of compounds are different. Ionic compounds are usually solids at room temperature, whereas covalently bonded compounds can be solids, liquids, or gases. There's more. Ionic compounds (salts) usually have a much higher melting point than covalent compounds. In addition, ionic compounds tend to be electrolytes, and covalent compounds tend to be nonelectrolytes. (Chapter 13 explains all about ionic bonds, electrolytes, and nonelectrolytes.)

I know just what you're thinking: "If metals react with nonmetals to form ionic bonds, and nonmetals react with other nonmetals to form covalent bonds, do metals react with other metals?" The answer is yes and no.

Metals don't really react with other metals to form compounds. Instead, metals combine to form *alloys,* solutions of one metal in another. But in a situation called *metallic bonding,* which affects both alloys and pure metals, the valence electrons of each metal atom are donated to an electron pool, commonly called a *sea of electrons,* and are shared by all the atoms in the metal. These valence electrons are free to move throughout the sample instead of being tightly bound to an individual metal nucleus. The ability of the valence electrons to flow throughout the entire metal sample is why metals tend to be conductors of electricity and heat and is responsible for the luster of metals.

Understanding multiple bonds

Covalent bonding is the sharing of one *or more* electron pairs. In hydrogen and most other diatomic molecules, only one electron pair is shared. But in many covalent bonding situations, more than one electron pair is shared. This section shows you an example of a molecule in which more than one electron pair is shared.

Energy is needed to break a covalent bond. The resistance of the bond to breaking is called its bond strength. In general, more energy is needed to break a double bond than a single bond if the same elements are involved. For example, a carbon-to-carbon double bond (two shared pairs of electrons) has a higher bond strength (requires more energy to break the bond) than a carbon-to-carbon single bond. The double bond isn't twice as strong as a single bond, but its strength is considerably greater. And a triple bond is stronger yet. Chemists also observe that multiple bonds are shorter in bond length (the distance between the nuclei of the bonded atoms) than single bonds — double bonds are shorter than single bonds, and triple bonds are shorter than double bonds.

Nitrogen (N_2) is a diatomic molecule in the *VA family* on the periodic table, meaning that it has five valence electrons (see Chapter 5 for a discussion of families on the periodic table). So nitrogen needs three more valence electrons to complete its octet. A nitrogen atom can fill its octet by sharing three electrons with another nitrogen atom, forming three covalent bonds, a triple bond. The triple-bond formation of nitrogen is shown in Figure 14-3.

A triple bond isn't quite three times as strong as a single bond, but it's a very strong bond. In fact, the triple bond in nitrogen is one of the strongest bonds known. This strong bond is what makes nitrogen gas very stable and resistant to reaction with other chemicals. It's also why many explosive compounds (such as TNT and ammonium nitrate) contain nitrogen. When these compounds break apart in a chemical reaction, nitrogen gas (N_2) is formed and a large amount of energy is released.

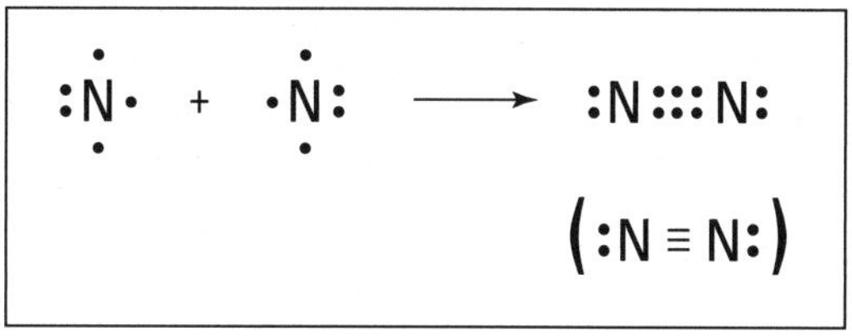

Figure 14-3:
Triple bond
formation
in N_2.

$$:N\cdot \ + \ \cdot N: \longrightarrow \ :N:::N:$$

$$\left(:N \equiv N:\right)$$

Carbon dioxide (CO_2) is another example of a compound containing a multiple bond. Carbon can react with oxygen to form carbon dioxide. Carbon has four valence electrons, and oxygen has six. Carbon can share two of its valence electrons with each of the two oxygen atoms, forming two double bonds. These double bonds are shown in Figure 14-4.

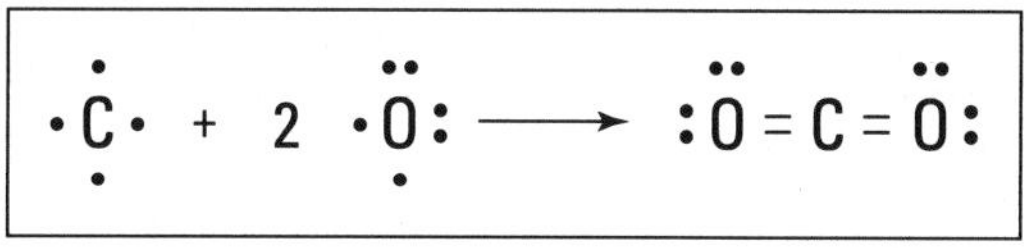

Figure 14-4:
Formation
of carbon
dioxide.

$$\cdot C\cdot \ + \ 2 \ \cdot O: \longrightarrow \ :O = C = O:$$

There are no salt molecules!

A *molecule* is a compound that's covalently bonded. Referring to sodium chloride, which has ionic bonds, as a molecule is technically incorrect, but lots of chemists (and chemistry students) do it anyway. The mistake is kind of like using the wrong fork at a formal dinner. Some people may notice, but most don't notice or don't care. But just so you know, the correct term for ionic compounds is *formula unit.*

Naming Binary Covalent Compounds

Binary compounds are compounds made up of only two elements, such as carbon dioxide (CO_2). Prefixes are used in the names of binary compounds to indicate the number of atoms of each nonmetal present, so being able to recognize binary compounds and apply the correct naming rules is important. Table 14-1 lists the most common prefixes for binary covalent compounds.

Table 14-1 Common Prefixes for Binary Covalent Compounds

Number of Atoms	Prefix
1	mono-
2	di-
3	tri-
4	tetra-
5	penta-
6	hexa-
7	hepta-
8	octa-
9	nona-
10	deca-

In general, the prefix *mono-* is rarely used. Carbon monoxide is one of the few compounds that uses it.

Take a look at the following examples to see how to use the prefixes when naming binary covalent compounds (I've bolded the prefixes for you):

CO_2 carbon **di**oxide

P_4O_{10} **tetra**phosphorus **dec**oxide (Chemists try to avoid putting an *a* and an *o* together with the oxide name, as in dec**a**oxide, so they normally drop the *a* off the prefix.)

SO_3 sulfur **tri**oxide

N_2O_4 **di**nitrogen **tetr**oxide

This naming system is used only with binary, nonmetal compounds, with one exception — MnO_2 is commonly called manganese dioxide.

Learning Many Formulas in a Little Time

In Chapter 13, I show you how to predict the formula of an ionic compound, based on the loss and gain of electrons, to reach a noble gas configuration. (For example, if you react Ca with Cl, you can predict the formula of the resulting salt — $CaCl_2$.) You really can't make that type of prediction with covalent compounds, because they can combine in many ways, and many different possible covalent compounds may result.

Most of the time, you have to know the formula of the molecule you're studying. But you may have several different types of formulas, and each gives a slightly different amount of information. Oh joy. Not to worry though. I provide the 411 on what you need to know about formulas in the following sections to make the prediction easier.

Empirical formula: Just the elements

The *empirical formula* indicates the different types of elements in a molecule and the lowest whole-number ratio of each kind of atom in the molecule. For example, suppose you have a compound with the empirical formula C_2H_6O. Three different kinds of atoms are in the compound, C, H, and O, and they're in the lowest whole-number ratio of 2 C to 6 H to 1 O. So the actual formula (called the *molecular formula* or *true formula*) may be C_2H_6O, $C_4H_{12}O_2$, $C_6H_{18}O_3$, $C_8H_{24}O_4$, or another multiple of 2:6:1.

Molecular or true formula: Inside the numbers

The *molecular formula,* or *true formula,* tells you the kinds of atoms in the compound and the actual number of each atom. You may determine, for example, that the empirical formula C_2H_6O is actually the molecular formula, too, meaning that the compound actually has two carbon atoms, six hydrogen atoms, and one oxygen atom.

For ionic compounds, this formula is enough to fully identify the compound, but it's not enough to identify covalent compounds. Look at the Lewis formulas presented in Figure 14-5. Both compounds have the molecular formula of C_2H_6O.

It's always important to KISS

A lot of molecules obey the octet rule: Each atom in the compound ends up with a full octet of eight electrons filling its valence energy level. However, like most rules, the octet rule does have exceptions. Some stable molecules have atoms with just 6 electrons, and some have 10 or 12. Take a peek at Chapter 15 for some specific examples of compounds that do not obey the octet rule. For the most part in this book, I concentrate on situations in which the octet rule is obeyed.

I pretty much stick to the KISS principle — Keep It Simple, Silly. Electron-dot formulas are used quite a bit by organic chemists in explaining why certain compounds react the way they do and are the first step in determining the molecular geometry of a compound (which I describe in Chapter 15).

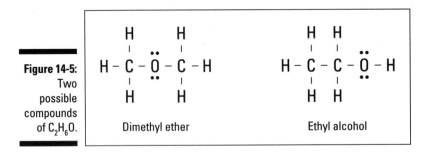

Figure 14-5:
Two possible compounds of C_2H_6O.

Dimethyl ether   Ethyl alcohol

Both compounds in Figure 14-5 have two carbon atoms, six hydrogen atoms, and one oxygen atom. The difference is in the way the atoms are bonded, or what's bonded to what, and it makes them two entirely different compounds with two entirely different sets of properties. The one on the left is called dimethyl ether. This compound is used in some refrigeration units and is highly flammable. The one on the right is ethyl alcohol, the drinking variety of alcohol. Simply knowing the molecular formula isn't enough to distinguish between the two compounds. Can you imagine going into a restaurant and ordering a shot of C_2H_6O and getting dimethyl ether instead of tequila?

Compounds that have the same molecular formula but different structures are called *isomers* of each other. To identify the *exact* covalent compound, you need its structural formula, which I discuss in the next section.

Structural formula: Add the bonding pattern

To write a formula that stands for the exact compound you have in mind, you often must write the structural formula instead of the molecular formula. The *structural formula* shows the elements in the compound, the exact number of each atom in the compound, and the bonding pattern for the compound. The electron-dot formula and Lewis formula are examples of structural formulas.

Writing the electron-dot formula for water

The following steps explain how to write the electron-dot formula for a simple molecule — water — and provide some general guidelines and rules to follow:

1. **Write a skeletal structure showing a reasonable bonding pattern using just the element symbols.**

   The skeletal structure involves only the atomic symbols and not the valence electrons. Often, most atoms are bonded to a single atom. This atom is called the *central atom*. Hydrogen and the halogens are very rarely, if ever, central atoms. Carbon, silicon, nitrogen, phosphorus, oxygen, and sulfur are always good candidates, because they form more than one covalent bond to fill their valence energy level. In the case of water, H_2O, oxygen is the central element and the hydrogen atoms are both bonded to it. The bonding pattern looks like this:

   O H
   H

   The hydrogen atoms can go anywhere around the oxygen. I put the hydrogen atoms at a 90-degree angle to each other, but it really doesn't matter when writing electron-dot (or Lewis) formulas.

2. **Take all the valence electrons from all the atoms and throw them into an electron pot.**

   Each hydrogen atom has 1 electron, and the oxygen atom has 6 valence electrons (VIA family), so you have 8 electrons in your electron pot. You use those electrons to make your bonds and complete each atom's octet.

   O H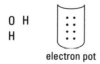
   H

   electron pot

3. **Use the $N - A = S$ equation to figure the number of bonds in this molecule. In this equation,**

   - N equals the sum of the number of valence electrons needed by each atom. N has only two possible values — 2 or 8. If the atom is hydrogen, it's 2; if it's anything else, it's 8.

   - A is the number of valence electrons in your electron pot — the sum of the number of valence electrons *available* for each atom. If you're doing the structure of an ion, you add one electron for every unit of negative charge if it's an anion or subtract one electron for every unit of positive charge if it's a cation.

   - S equals the number of electrons *shared* in the molecule. And if you divide S by 2, you have the number of covalent bonds in the molecule.

   So in the case of water,

   - $N = 8 + 2(2) = 12$ (8 valence electrons for the oxygen atom, plus 2 each for the two hydrogen atoms)

   - $A = 6 + 2(1) = 8$ (6 valence electrons for the oxygen atom, plus 1 for each of the two hydrogen atoms)

   - $S = 12 - 8 = 4$ (four electrons shared in water), and $S/2 = 4/2 = 2$ bonds

   You now know that water molecules have two bonds (two shared pairs of electrons).

4. **Distribute the electrons from your electron pot to account for the bonds.**

   You use 4 electrons from the 8 in the pot, which leaves you with 4 to distribute later. At least one bond must connect your central atom to the atoms surrounding it.

   electron pot

5. **Distribute the rest of the electrons (normally in pairs) so that each atom achieves its full octet of electrons.**

   Remember that hydrogen needs only 2 electrons to fill its valence energy level. In this case, each hydrogen atom has 2 electrons, but the oxygen atom has only 4 electrons, so the remaining 4 electrons are placed around the oxygen, emptying your electron pot. The completed electron-dot formula for water is shown in Figure 14-6.

Figure 14-6:
Electron-dot
formula
of H_2O.

Notice that this structural formula actually shows two types of electrons: *bonding electrons,* the electrons that are shared between two atoms, and *non-bonding electrons,* the electrons that are not being shared. The last 4 electrons (2 electron pairs) that you put around oxygen are not being shared, so they're nonbonding electrons.

Writing the Lewis formula for water

If you want the Lewis formula for water, all you have to do is substitute a dash for every bonding pair of electrons. This structural formula is shown in Figure 14-7.

Figure 14-7:
The Lewis
formula
for H_2O.

Writing the Lewis formula for C_2H_4O

Here's an example of a Lewis formula that's a little more complicated — C_2H_4O.

The compound has the following framework:

H
H C C O
H H

electron pot

Notice that the compound has not 1 but 2 central atoms — the 2 carbon atoms. You can put 18 valence electrons into the electron pot: 4 for each carbon atom, 1 for each hydrogen atom, and 6 for the oxygen atom.

Now apply the $N - A = S$ equation:

$N = 2(8) + 4(2) + 8 = 32$ (2 carbon atoms with 8 valence electrons each, plus 4 hydrogen atoms with 2 valence electrons each, plus an oxygen atom with 8 electrons)

$A = 2(4) + 4(1) + 6 = 18$ (4 electrons for each of the two carbon atoms, plus 1 electron for each of the 4 hydrogen atoms, plus 6 valence electrons for the oxygen atom)

$S = 32 - 18 = 14$, and $S/2 = 14/2 = 7$ bonds

Add single bonds between the carbon atoms and the hydrogen atom, between the 2 carbon atoms, and between the carbon atom and oxygen atom. That's 6 of your 7 bonds.

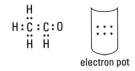

electron pot

The seventh bond can go only one place, and that's between the carbon atom and the oxygen atom. It can't be between a carbon atom and a hydrogen atom, because that would overfill hydrogen's valence energy level. And it can't be between the two carbon atoms, because that would give the carbon on the left 10 electrons instead of 8. So there must be a double bond between the carbon atom and the oxygen atom. The 4 remaining electrons in the pot must be distributed around the oxygen atom, because all the other atoms have reached their octet. The electron-dot formula is shown in Figure 14-8.

Figure 14-8:
Electron-dot
formula
of C_2H_4O.

$$H:C:C::O$$

If you convert the bonding pairs to dashes, you have the Lewis formula of C_2H_4O, as shown in Figure 14-9.

Figure 14-9:
The Lewis
formula
for C_2H_4O.

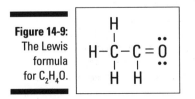

I like the Lewis formula because it enables you to show a lot of information without having to write all those little dots. But it, too, is rather bulky. Sometimes chemists (who are, in general, a lazy lot) use *condensed structural formulas* to show bonding patterns. They may condense the Lewis formula by omitting the nonbonding electrons and grouping atoms together and/or by omitting certain dashes (covalent bonds). A couple of condensed formulas for C_2H_4O are shown in Figure 14-10.

Figure 14-10:
Condensed
structural
formulas
for C_2H_4O.

$$CH_3 - CH = O$$

$$CH_3CHO$$

Sharing Electron Pairs — Sometimes Equally and Sometimes Not

One of my favorite lines from the book *Animal Farm* is "All animals are equal, but some animals are more equal than others." The same is true of covalent bonds — electron pairs may be shared, but not always equally.

When a chlorine atom covalently bonds to another chlorine atom, the shared electron pair is shared equally. The electron density that comprises the covalent bond is located halfway between the two atoms. Each atom attracts the two bonding electrons equally because each nuclei has the same number of protons.

But what happens when the two atoms involved in a bond aren't the same? The two positively charged nuclei have different attractive forces; they "pull" on the electron pair to different degrees. The end result is that the electron pair is shifted toward one atom. But the question is, "Which atom does the electron pair shift toward?" Electronegativities provide the answer.

Attracting electrons: Electronegativities

Electronegativity is the strength an atom has to attract a bonding pair of electrons to itself. The larger the value of the electronegativity, the greater the atom's strength to attract a bonding pair of electrons. Figure 14-11 shows the electronegativity values of the various elements below each element symbol on the periodic table. Notice that, with a few exceptions, the electronegativities increase from left to right, in a period, and decrease from top to bottom, in a family.

Electronegativities of the Elements

Increasing →

Decreasing →

1 H 2.1																
3 Li 1.0	4 Be 1.5											5 B 2.0	6 C 2.5	7 N 3.0	8 O 3.5	9 F 4.0
11 Na 0.9	12 Mg 1.2											13 Al 1.5	14 Si 1.8	15 P 2.1	16 S 2.5	17 Cl 3.0
19 K 0.8	20 Ca 1.0	21 Sc 1.3	22 Ti 1.5	23 V 1.6	24 Cr 1.6	25 Mn 1.5	26 Fe 1.8	27 Co 1.9	28 Ni 1.9	29 Cu 1.9	30 Zn 1.6	31 Ga 1.6	32 Ge 1.8	33 As 2.0	34 Se 2.4	35 Br 2.8
37 Rb 0.8	38 Sr 1.0	39 Y 1.2	40 Zr 1.4	41 Nb 1.6	42 Mo 1.8	43 Tc 1.9	44 Ru 2.2	45 Rh 2.2	46 Pd 2.2	47 Ag 1.9	48 Cd 1.7	49 In 1.7	50 Sn 1.8	51 Sb 1.9	52 Te 2.1	53 I 2.5
55 Cs 0.7	56 Ba 0.9	57 La 1.1	72 Hf 1.3	73 Ta 1.5	74 W 1.7	75 Re 1.9	76 Os 2.2	77 Ir 2.2	78 Pt 2.2	79 Au 2.4	80 Hg 1.9	81 Tl 1.8	82 Pb 1.9	83 Bi 1.9	84 Po 2.0	85 At 2.2
87 Fr 0.7	88 Ra 0.9	89 Ac 1.1														

Figure 14-11: Electronegativities of the elements.

Electronegativities are useful because they give information about what will happen to the bonding pair of electrons when two atoms bond. Basically three types of bonds can be formed.

✔ **Nonpolar covalent bond:** This bond has an electron pair that's equally shared. You have a nonpolar covalent bond anytime the two atoms involved in the bond are the same or anytime the difference in the electronegativities of the atoms involved in the bond is very small. For example, look at the Cl_2 molecule. Chlorine has an electronegativity value of 3.0, as shown in Figure 14-11. Each chlorine atom attracts the bonding electrons with a force of 3.0. Because there's an equal attraction, the bonding electron pair is shared equally between the two chlorine atoms and is located halfway between the two atoms.

✔ **Polar covalent bond:** The electron pair in this bond is shifted toward one atom. The atom that more strongly attracts the bonding electron pair is slightly more negative, while the other atom is slightly more positive. The larger the difference in the electronegativities, the more negative and positive the atoms become.

Consider hydrogen chloride (HCl). Hydrogen has an electronegativity of 2.1, and chlorine has an electronegativity of 3.0. The electron pair that is bonding HCl together shifts toward the chlorine atom because it has a larger electronegativity value. Check out the next section for more information on these types of bonds.

✔ **Ionic bond:** In this case, the bonding electrons are totally removed from one of the atoms and ions are formed. Now look at a case in which the two atoms have extremely different electronegativities — sodium chloride (NaCl). Sodium chloride is ionically bonded (see Chapter 13 for information on ionic bonds). An electron has transferred from sodium to chlorine. Sodium has an electronegativity of 1.0, and chlorine has an electronegativity of 3.0. That's an electronegativity difference of 2.0 (3.0 – 1.0), making the bond between the two atoms very, very polar. In fact, the electronegativity difference provides another way of predicting the kind of bond that will form between two elements.

The following table breaks down these three types of bonds that are formed and shows their electronegativity difference:

Electronegativity Difference	*Type of Bond Formed*
0.0 to 0.2	Nonpolar covalent
0.3 to 1.4	Polar covalent
> 1.5	Ionic

The presence of a polar covalent bond in a molecule can have some pretty dramatic effects on the properties of a molecule.

Polar covalent bonding

If the two atoms involved in the covalent bond are not the same, the bonding pair of electrons are pulled toward one atom, with that atom taking on a slight (partial) negative charge and the other atom taking on a partial positive charge. In many cases, the molecule then has a positive end and a negative end, and can be referred to as a *dipole* (think of a magnet). Figure 14-12 shows a couple of examples of molecules in which dipoles have formed. (The little Greek symbol by the charges refers to a *partial* charge.)

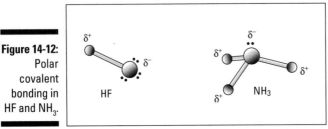

Figure 14-12:
Polar covalent bonding in HF and NH₃.

In hydrogen fluoride (HF), the bonding electron pair is pulled much closer to the fluorine atom than to the hydrogen atom, so the fluorine end becomes partially negatively charged and the hydrogen end becomes partially positively charged. The same thing takes place in ammonia (NH₃); the nitrogen has a greater electronegativity than hydrogen, so the bonding pairs of electrons are more attracted to it than to the hydrogen atoms. The nitrogen atom takes on a partial negative charge, and the hydrogen atoms take on a partial positive charge.

The presence of a polar covalent bond explains why some substances act the way they do in a chemical reaction: Because this type of molecule has a positive end and a negative end, it can attract the part of another molecule with the opposite charge.

In addition, this type of molecule can act as a weak electrolyte because a polar covalent bond allows the substance to act as a conductor. So if a chemist wants a material to act as a good *insulator* (a device used to separate conductors), the chemist looks for a material with as weak a polar covalent bond as possible.

Wondering about water: A really strange molecule

Water (H₂O) has some very strange chemical and physical properties. It can exist in all three states of matter at the same time. Imagine that you're sitting in your hot tub (filled with *liquid* water) watching the steam *(gas)* rise from the surface as you enjoy a cold drink from a glass filled with ice *(solid)* cubes. Very few other chemical substances can exist in all these physical states in this close of a temperature range.

And those ice cubes are floating! In the solid state, the particles of matter are usually much closer together than they are in the liquid state. So if you put a solid into its corresponding liquid, it sinks — but not for water. Its solid state is less dense than its liquid state, so it floats. Imagine what would happen if ice sank. In the winter, the lakes would freeze, and the ice would sink to the bottom, exposing more water. The extra exposed water would then freeze and sink, and so on, until the entire lake was frozen solid. This would destroy the aquatic life in the lake in no time. Fortunately, instead, the ice floats and insulates the water underneath it, protecting aquatic life. And water's boiling point is unusually high. Other compounds similar in weight to water have a *much* lower boiling point.

Another unique property of water is its ability to dissolve a large variety of chemical substances. It dissolves salts and other ionic compounds, as well as polar covalent compounds such as alcohols and organic acids. In fact, water is sometimes called the universal solvent because it can dissolve so many things. It can also absorb a large amount of heat, which allows large bodies of water to help moderate the temperature on earth.

Water has many unusual properties because of its polar covalent bonds. Oxygen has a larger electronegativity than hydrogen, so the electron pairs are pulled in closer to the oxygen atom, giving it a partial negative charge. Subsequently, both of the hydrogen atoms take on a partial positive charge. The partial charges on the atoms created by the polar covalent bonds in water are shown in Figure 14-13.

Figure 14-13:
Polar
covalent
bonding in
water.

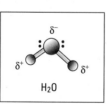

H_2O

Water is a dipole and acts like a magnet, with the oxygen end having a negative charge and the hydrogen end having a positive charge. These charged ends can attract other water molecules. The partially negatively charged oxygen atom of one water molecule can attract the partially positively charged hydrogen atom of another water molecule. This attraction between the molecules occurs frequently and is a type of *intermolecular force* (force between different molecules).

Intermolecular forces can be of three different types:

- ✔ **London force:** Also called the *dispersion force,* this force is a very weak type of attraction that generally occurs between nonpolar covalent molecules, such as nitrogen (N_2), hydrogen (H_2), or methane (CH_4). It results from the ebb and flow of the electron orbitals, giving a very weak and very brief charge separation around the bond.

- ✔ **Dipole-dipole interaction:** This intermolecular force occurs when the positive end of one dipole molecule is attracted to the negative end of another dipole molecule. It's much stronger than a London force, but it's still pretty weak.

- ✔ **Hydrogen bond:** The third type of interaction is really just an extremely strong dipole-dipole interaction that occurs when a hydrogen atom on one molecule is bonded to one of three extremely electronegative elements — O, N, or F — on another molecule. These three elements have a very strong attraction for the bonding pair of electrons, so the atoms involved in the bond take on a large amount of partial charge. This bond turns out to be highly polar — and the higher the polarity, the more effective the bond. It's only about 5 percent of the strength of an ordinary covalent bond, but still very strong for an intermolecular force. The hydrogen bond is the type of interaction that's present in water (see Figure 14-14). For a more complete discussion of intermolecular forces, check out Chapter 17.

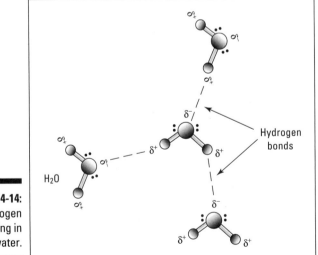

Figure 14-14:
Hydrogen
bonding in
water.

Water molecules are stabilized by these hydrogen bonds, so breaking up (separating) the molecules is very hard. The hydrogen bonds account for water's high boiling point and ability to absorb heat. When water freezes, the hydrogen bonds lock water into an open framework that includes a lot of empty space. In liquid water, the molecules can get a little closer to each other, but when the solid forms, the hydrogen bonds result in a structure that contains large holes. The holes increase the volume and decrease the density. This process explains why the density of ice is less than that of liquid water (the reason ice floats). The structure of ice is shown in Figure 14-15, with the hydrogen bonds indicated by dotted lines.

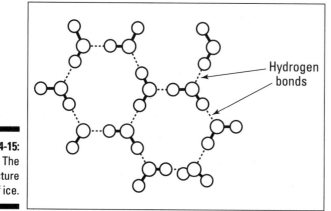

Hydrogen bonds

Figure 14-15:
The
structure
of ice.

Chapter 15

What Do Molecules Really Look Like? Molecular Geometry and Hybridization

. . . .

In This Chapter

▶ Sizing up the importance of shape

▶ Delving into electron and molecular geometry

▶ Finding out what the valence bond theory reveals about covalent bonds

▶ Bonding with the molecular orbital theory

. . . .

*A*toms come together to make compounds with ionic bonding or covalent bonding. Ionic compounds are held together in a type of crystal lattice. You can describe the shape of the lattice but not the individual ion. In covalent bonding, the atoms combine to make a molecule, which has a shape that can be described. This chapter is about determining the shape of molecules created by covalent bonding.

The shape of a molecule can be predicted in two ways:

> ✔ **The VSEPR method:** This method, which stands for *valence-shell electron-pair repulsion,* relies on the Lewis structure of the molecule.
>
> ✔ **The valence bond method:** This approach relies on the electron configuration of the central atom.

So why use two methods when they both give the same result? The VSEPR method tends to be simpler to apply, but the valence bond approach allows you to determine and rationalize other properties of molecules. I discuss both methods of determining shape in this chapter. I also throw in a discussion on the molecular orbital theory, which can also be used to predict the shape of molecules, for no additional charge.

Seeing How Shape Matters

The *shape* (orientation in space of the atoms in a molecule) of a molecule may determine a great deal of its properties. For example, the shape of many organic molecules determines whether they're reactive in certain circumstances. The shape is especially indicative of reactivity for *enzymes,* biological catalysts, that are found in the human body. If the shape of a particular enzyme is altered, that enzyme becomes useless in helping a certain biochemical reaction occur. Shape is also important in the complex molecules of drugs; it makes the drugs active but also leads to side effects. The following sections examine the importance of shape and the role polarity plays in an atom's shape.

Getting charged with polarity

One of the important consequences of the shape of a molecule is the *polarity* (separation of charge) of the molecule. Some molecules are *polar;* they have charged regions that can attract oppositely charged regions in other molecules. Other molecules are *nonpolar;* they have no charged regions to attract other molecules. Whether or not a molecule is polar is related to the presence of charge separation in the bonds and also on the shape of the molecule. You can have polar bonds, but if they're symmetrically distributed in the molecule, the molecule itself won't be polar.

A polar molecule has a partially positive charge at one end and a partially negative charge at the opposite end. The partial charge on one molecule attracts the opposite partial charge on another molecule. This attraction is less than the attraction between ions because these molecules only have partial (slight) charges. This attraction, even if less than in an ionic situation, is a significant factor in the physical properties of these molecules.

To explore an example of polarity, start by looking at Figure 15-1, which shows the Lewis structures of two diatomic molecules, the hydrogen molecule, H_2, and the chlorine fluoride molecule, ClF. (The Lewis structure is a way of showing the structural formula of a molecule; refer to Chapter 14 for more information.)

Figure 15-1:
Lewis
structures
of hydrogen
and chlorine
fluoride.

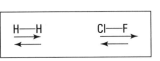

The arrows beneath the two structures illustrate the relative electronegativities of the elements. In the hydrogen molecule, the two arrows are the same length because both atoms have equal electronegativities. In the chlorine fluoride molecule, the arrows are of different lengths because the electronegativities are different. The electronegativity of fluorine is greater than that of chlorine, so the arrow pointing to the fluorine is longer than the arrow pointing to the chlorine.

If you combine the two equal arrows in the hydrogen molecule, they cancel. Hydrogen atoms, due to their identical electronegativities, pull equally on the shared electrons, resulting in an equal sharing of the electrons. Therefore, H_2 is nonpolar, because the charges on either end of the molecule are the same and equal.

However, if you combine the unequal arrows in chlorine fluoride, you're left with a net arrow pointing toward the fluorine end of the molecule. This remaining arrow indicates that the shared negative electrons are pulled closer toward the fluorine. The unequal sharing of the electrons leads to a slight (partial) negative charge on the fluorine, leaving a slight (partial) positive charge on the chlorine. Thus, ClF is a polar molecule. (If the difference in electronegativities were greater, the more electronegative element would strip the electron from the less electronegative element to form an ionic bond.)

A lowercase Greek letter delta, δ, indicates the presence of a partial charge. If you add the partial charges to the picture of the chlorine fluoride molecule, you get the structure shown in Figure 15-2.

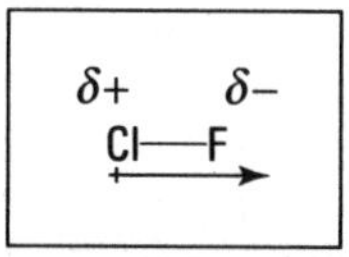

Figure 15-2:
The Lewis structure of chlorine fluoride shows partial charges.

The arrow below the molecule is an alternative to the partial charges. The arrow has a cross at the positive end and points toward the more negative end.

Predicting polarity

Predicting the polarity of diatomic molecules is simple. A diatomic molecule is polar if the bond is a polar covalent bond, and it is nonpolar if the covalent bond is nonpolar. Thus, molecules such as HCl, NO, and ClF are polar, and molecules such as H_2, Cl_2, and N_2 are nonpolar. Refer to Chapter 14 for a discussion of electronegativities and polar covalent bonding.

A polar molecule is a dipole. *Dipole* refers to the two poles, the partially positive pole and the partially negative pole. The magnitude of the dipole is related to the difference in the electronegativities (and, to a lesser degree, to the distance between the two atoms). The *dipole moment* is a measure of the magnitude of the dipole. Nonpolar molecules have a zero dipole moment, and all polar molecules have a nonzero dipole moment.

Although the prediction of the polarity of a diatomic molecule is relatively simple, what happens when there are more than two atoms? I use water to illustrate this problem because water clearly demonstrates that, if you want to know about the polarity of a molecule, you must first know the *molecular geometry,* the arrangement of the atoms and electrons in three-dimensional space. Figure 15-3 shows the correct Lewis structures for the water molecule.

Figure 15-3:
Two
possible
Lewis
structures
of water.

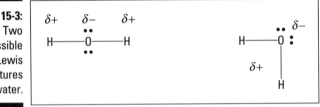

Due to the difference in electronegativities between the hydrogen and the oxygen, all the bonds are polar covalent. The oxygen atom, being more electronegative than the hydrogens, has a partial negative charge, leaving a partial positive charge on each hydrogen atom. The linear arrangement in the Lewis structure on the left has a partial positive charge at each end and a partial negative charge in the center. Because the molecule does not have a partial positive end and a partial negative end, this molecule is nonpolar. The bent molecule, however, is polar because it has a partial negative oxygen on one end of the molecule and two hydrogens with partial positives at the other end.

The properties of water indicate which structure is more likely. A nonpolar water molecule would have properties similar to other nonpolar molecules of approximately the same molecular mass, such as methane. For example, it should boil at about –100 degrees Celsius. Because water boils 200 degrees higher (+100 degrees Celsius), you can reasonably assume that water must be polar instead of nonpolar.

Delving Into Electron and Molecular Geometry (VSEPR)

One method to predict the shape of molecules is the *valence-shell electron-pair repulsion (VSEPR) theory*. The basis of this theory is that the valence-shell electron pairs around a central atom try to move as far away from each other as possible (like two supermodels showing up at a party wearing the same dress). Electrons spread out to minimize the repulsion between their like (negative) charges. This theory includes both electrons in bonds and nonbonding or lone-pair electrons.

With this VSEPR method, I show you how to actually determine two geometries. The two geometries are as follows:

- ✔ **Electron-pair geometry:** Sometimes called the *electron-group geometry*, it considers all electron pairs surrounding a nucleus.
- ✔ **Molecular geometry:** The nonbonding electrons become "invisible," and you consider only the arrangement of the atomic nuclei.

In determining geometry, double and triple bonds count the same as single bonds.

In order to determine the electron-group and molecular geometry, follow these steps:

1. **Write the Lewis electron-dot formula of the compound.**

   Refer to Chapter 14 for the rules of writing Lewis structures.

2. **Determine the number of electron pair groups surrounding the central atom(s).**

   Remember that double and triple bonds count the same as a single bond.

3. **Determine the geometric shape that maximizes the distance between the electron groups.**

   Check out Table 15-1 for electron-pair geometry, the shapes associated with the number of electron pairs.

4. **Mentally allow the nonbonding electrons to become invisible.**

   They are still present and are still repelling the other electron pairs. However, you just don't "see" them.

5. **Determine the molecular geometry from the arrangement of bonding pairs around the central atom by referring to Table 15-1.**

Table 15-1		Predicting Molecular Shape with the VSEPR Theory	
Total Number of Electron Pairs	Number of Bonding Pairs	Electron-Pair Geometry	Molecular Geometry
2	2	Linear	Linear
3	3	Trigonal planar	Trigonal planar
3	2	Trigonal planar	Bent, V shaped
3	1	Trigonal planar	Linear
4	4	Tetrahedral	Tetrahedral
4	3	Tetrahedral	Trigonal pyramidal
4	2	Tetrahedral	Bent, V shaped
5	5	Trigonal bipyramidal	Trigonal bipyramidal
5	4	Trigonal bipyramidal	Seesaw
5	3	Trigonal bipyramidal	T shaped
5	2	Trigonal bipyramidal	Linear
6	6	Octahedral	Octahedral
6	5	Octahedral	Square pyramidal
6	4	Octahedral	Square planar

Even though you normally don't have to worry about more than four electron pairs around the central atom (because of the octet rule), I put some of the less common exceptions to the octet rule in Table 15-1. Figure 15-4 shows some of the more common shapes mentioned in the table.

Determine the shapes of water (H_2O) and ammonia (NH_3).

To do so, the first thing you have to do is determine the Lewis formula for each compound. Follow the rules outlined in Chapter 14 and write the Lewis formulas as shown in Figure 15-5.

For water, four electron pairs are around the oxygen atom, so the electron-pair geometry is *tetrahedral* (refer to Figure 15-4). Only two of these four electron pairs are involved in bonding, so the molecular shape is bent or V shaped. Because the molecular shape for water is V shaped, I always show water with the hydrogen atoms at about a 90-degree angle to each other — it's a good approximation of the actual shape.

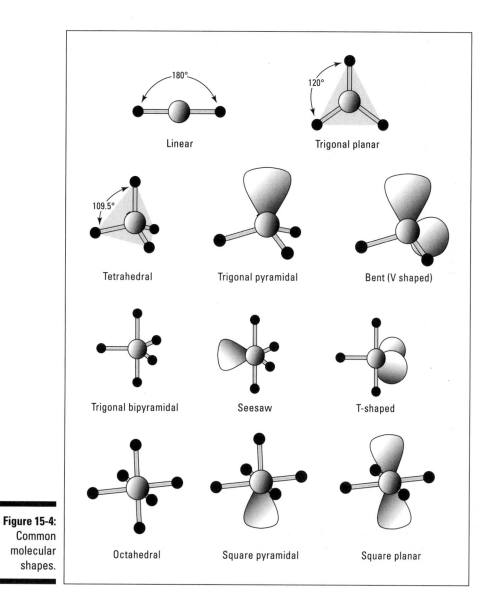

Figure 15-4: Common molecular shapes.

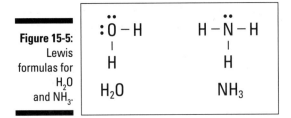

Figure 15-5:
Lewis
formulas for
H_2O
and NH_3.

Ammonia also has four electron pairs around the nitrogen central atom, so its electron-pair geometry is tetrahedral as well. Only one of the four electron pairs is nonbonding, however, so its molecular shape is *trigonal pyramidal*. This shape is like a three-legged milk stool, with the nitrogen being the seat, the three bonding pairs of electrons being the legs, and the "invisible" lone pair of nonbonding electrons sticking straight up from the seat. You'd get a surprise if you sat on an ammonia stool!

Comprehending the Valence Bond Theory (Hybridization)

Another method to determine molecular geometry involves using the valence bond theory. *Valence bond theory* explains covalent bonding in terms of the blending of atomic orbitals to form new types of orbitals, hybrid orbitals. *Hybrid orbitals* are orbitals formed when the atomic orbitals of the central atom combine when forming a compound. However, the total number of orbitals doesn't change; the number of hybrid orbitals equals the number of atomic orbitals used. The number and type of atomic orbitals used determine what type of hybrid orbitals form. Figure 15-6 shows the hybrid orbitals resulting from the mixing of s, p, and d orbitals. The atoms share electrons through the overlapping of their orbitals.

	Linear	Trigonal planar	Tetrahedral	Trigonal bipyramidal	Octahedral
Atomic orbitals mixed	one s one p	one s two p	one s three p	one s three p one d	one s three p two d
Hybrid orbitals formed	two sp	three sp^2	four sp^3	five sp^2d	six sp^3d^2
Unhybridized orbitals remaining	two p	one p	none	four d	three d
Orientation					

Figure 15-6:
Hybridization
involving
the s, p, and
d orbitals.

The type of hybridization formed depends on the type and number of atomic orbitals that are involved, and this in turn affects the shape of the resulting molecule:

- ✔ *sp hybridization* results from the overlap of one s orbital with one p orbital. Two sp-hybrid orbitals form with a bond angle of 180 degrees, which is called a *linear orientation.*

- ✔ *sp² hybridization* results from the overlap of one s orbital with two p orbitals. Three sp²-hybrid orbitals form with a trigonal planar orientation and a bond angle of 120 degrees. This type of bonding occurs in the formation of the C-to-C double bond, as in $CH_2=CH_2$.

- ✔ *sp³ hybridization* results from the combination of one s orbital and three p orbitals, resulting in four sp³-hybrid orbitals with a tetrahedral geometric orientation. You find this sp³ hybridization in carbon when it forms four single bonds.

- ✔ *sp³d hybridization* results from the blending of one s orbital, three p orbitals, and one d orbital. The result is five sp³d orbitals with a trigonal bipyramidal orientation. This type of bonding occurs in compounds like PCl_5, an exception to the octet rule.

- ✔ *sp³d² hybridization* occurs when one s orbital, three p orbitals, and two d orbitals come together to create an octahedral arrangement. SF_6 is an example. As you have seen before, SF_6 is an exception to the octet rule. If one of the bonding pairs in an sp³d² hybridization becomes a lone pair, a square pyramidal shape results; two lone pairs results in a square planar shape.

Figure 15-7 shows the hybridization found in ethylene, $H_2C=CH_2$. Each carbon has sp² hybridization. On each carbon, two of the hybrid orbitals overlap with an s orbital on a hydrogen atom to form a carbon-to-hydrogen covalent bond. The third sp²-hybrid orbital overlaps with the sp² hybrid on the other carbon to form a carbon-to-carbon covalent bond. Note that each carbon has a remaining p orbital that has not undergone hybridization. These p orbitals also overlap above and below a line joining the carbons.

Figure 15-7:
Hybridization in ethylene.

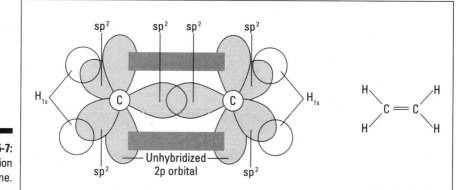

Ethylene has two types of bonds. *Sigma (σ) bonds* have the overlap of the orbitals on a line between the two atoms involved in the covalent bond. In ethylene, the C-H bonds and one of the C-C bonds are sigma bonds. *Pi (π) bonds* have the overlap of orbitals above and below a line through the two nuclei of the atoms involved in the bond. A double bond is always composed of one sigma bond and one pi bond. A carbon-to-carbon triple bond results from the overlap of one sp-hybrid orbital and two p orbitals on one carbon and the same overlap on the other carbon, resulting in one sigma bond (overlap of the sp-hybrid orbitals) and two pi bonds (overlap of two sets of p orbitals).

In a multiple bond, one of the bonds must always be a sigma bond and the others are pi.

Breaking Down the Molecular Orbital (MO) Theory

Another covalent bonding model is molecular orbital theory. In *molecular orbital (MO) theory,* atomic orbitals on the individual atoms combine to form molecular orbitals (MOs), which aren't hybrid orbitals. A molecular orbital covers the entire molecule and has a definite shape and energy. The combination of two atomic orbitals produces two molecular orbitals. (Like in hybridization, the total number of orbitals never changes.)

  ✔ **Bonding molecular orbital:** The bonding MO has a lower energy than the original atomic orbitals.

  ✔ **Antibonding molecular orbital:** The antibonding molecular orbital has a higher energy than the original atomic orbitals.

As shown in Figure 15-8, lower-energy orbitals are more stable than higher-energy orbitals. Notice that the bonding molecular orbital, because it's of lower energy than the original atomic orbitals, strengthens the bond.

Figure 15-8:
Formation of
bonding and
antibonding
molecular
orbitals
from the
overlap of
two atomic
orbitals.

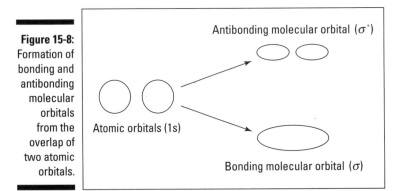

Antibonding molecular orbital (σ^*)

Atomic orbitals (1s)

Bonding molecular orbital (σ)

The end-to-end overlap of two p orbitals yields a σ bonding and a σ* antibonding molecular orbital. The side-by-side overlapping of p orbitals yields a π bond composed of one π bonding molecular orbital and one π* antibonding molecular orbital, as you can see in Figure 15-9.

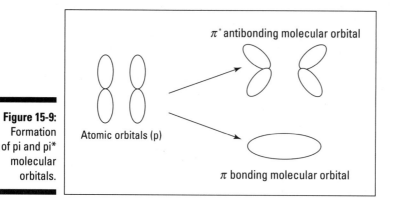

Figure 15-9: Formation of pi and pi* molecular orbitals.

After the molecular orbitals form, you put electrons in. You add electrons using the same rules you used for electron configurations (see Chapter 12).

- ✔ The lower-energy orbitals fill first.
- ✔ Each orbital has a maximum of two electrons.
- ✔ Orbitals of equal energy half-fill orbitals before pairing electrons.
- ✔ When two s atomic orbitals combine, two sigma (σ) molecular orbitals form. One is sigma bonding (σ), and the other is sigma antibonding (σ*).

Figure 15-10 shows the molecular orbital diagram for H_2.

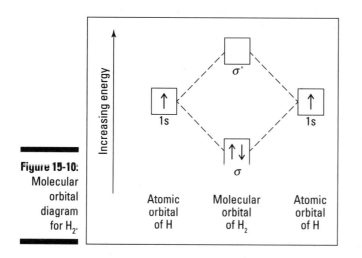

Figure 15-10: Molecular orbital diagram for H_2.

Notice that the two electrons (one from each hydrogen atom) have both gone into the sigma bonding MO. You can determine the bonding situation in the molecular orbital theory by calculating the MO bond order. The *MO bond order* is the number of electrons in bonding MOs minus the number of electrons in antibonding MOs, divided by 2. A stable bonding situation exists between two atoms when the bond order is greater than zero. The larger the bond order, the stronger the bond. For H_2 in Figure 15-10, the bond order is $(2-0)/2 = 1$.

When two sets of p orbitals combine, one sigma bonding and one sigma antibonding MO are formed along with two pi bonding MOs and two pi antibonding MOs. Figure 15-11 shows the MO diagram for O_2. For the sake of simplicity, I don't show the 1s orbitals of each oxygen or MOs here, just the valence-electron orbitals.

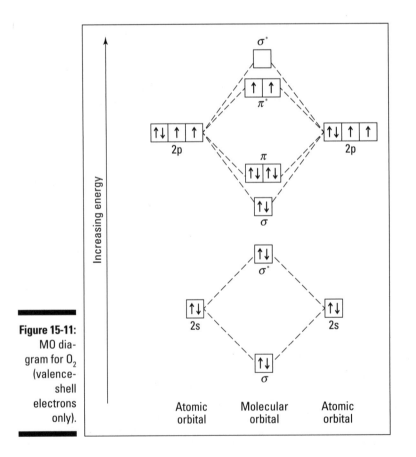

Figure 15-11: MO diagram for O_2 (valence-shell electrons only).

The bond order for O_2 is $(10-6)/2 = 2$. (Don't forget to count the bonding and antibonding electrons at energy level 1.)

Chapter 16

Tackling Periodic Trends

. . . .

. . . .

*I*n Chapter 5, I explain a lot of information concerning the periodic
table. I show you a number of ways to classify elements on the periodic
table — by groups or families (the vertical columns); by periods (the
horizontal rows); by metals, nonmetals and metalloids; and so on. I also
explain how the arrangement of electrons determines an element's position
on the periodic table and how those outermost electrons, the valence
electrons, determine the chemical properties of the elements.

In this chapter, I show you some trends in the properties of elements on
the periodic table. I know, when you think of trends you think of fashion —
bell-bottom pants and platform shoes. (Oops, did I just give you a clue as to
my age?) The periodic table has trends, too, and although they may not be
very stylish, these periodic trends can be useful in predicting speeds of
reaction, size of atoms and ions, and other characteristics. So sit back and
get trendy.

Checking Out the Importance of Size

When working with atoms and their parts, understanding the different sizes
and their relationship to each other is essential. The occupied energy levels
of the atom basically determine the atom's size. The nucleus is tiny; the
electrons in their energy levels around the nucleus really determine the
boundaries of the atom. Some additional factors, especially the effective
nuclear charge, have an effect on the electron and thus on the size of the
atom. I discuss these factors in the following sections.

Comprehending effective nuclear charge

The nucleus of an atom contains protons and neutrons. As I discuss in Chapter 4, the number of protons in a nucleus is the *atomic number.* Neutrons have no charge, whereas protons have a positive charge. If you add together the positive charges of all those protons, you have what's called the *nuclear charge.* This positive charge in the nucleus attracts electrons, which are negatively charged. (Unlike charges attract, and like charges repel.) If the atom contains only a single electron, the attraction force is straightforward. However, when more than one electron is present, the situation becomes more complicated.

The negative charge of one electron interferes with the attractive force of the nucleus to other electrons. Electrons occupy "energy levels" that form a sort of layered shell around the nucleus. When electrons closer to the nucleus come between another electron and the nucleus, they cause a greater interference. Therefore, electrons that are contained in the same energy level produce a minimum interference on each other, but the core electrons (inner-shell electrons) partially shield the outer electrons from the attraction of the positive charge in the nucleus, the nuclear charge. This shielding affects all electrons in an atom, but for trends in size, the shielding of the valence electrons (outermost electrons) is the key. Those outermost electrons are contained in the orbital with the highest value of the principal quantum number, *n,* and are the electrons that determine the size of the atomic radius. If shielding lessens the attractive force of the nucleus to the valence electrons, they're free to move a little bit farther from the nucleus, and the atomic diameter increases.

The core electrons screen the nuclear charge from the valence electrons. The actual charge attracting the valence electrons is called the effective nuclear charge. The *effective nuclear charge* is the nuclear charge minus the shielding effect of the core electrons. The shielding effect is essentially the same for any period on the periodic table, but when going down a family (group), the effect increases as you move toward the bottom of the column. The greater the shielding, the less the effective nuclear charge, allowing the valence electrons to move a little farther away from the nucleus. This increases the size of the atom.

Explaining changes in atomic radii

The atomic *radius* is the distance from the center of the nucleus to the valence energy level. It's how chemists describe the size of the atom or ion. It's important in determining the type of crystal lattice formed, solubility, and so on. On the periodic table, atomic *radii* (the plural of radius) increase from top to bottom within a family and decrease from left to right within a period.

When you move down a family, the atomic radii of the atoms increase, making each atom significantly larger than the one above it. This increase in size is the result of electrons occupying energy levels of increasing distance from the nucleus, increasing n values. The nuclear charge increases as you go down a family (increased numbers of protons in the nucleus), but an increasing number of core electrons shield the valence electrons from the charge in the nucleus. The effective nuclear charge on the valence electrons is thus decreased, allowing the electrons to move slightly away from the nucleus and countering the nuclear charge. Therefore, simply adding an energy level predominates in determining the size of the atomic radii.

Understanding the size changes as you move across a period isn't so simple. Unlike atoms in families, which add energy levels, atoms in periods add electrons to the same energy level. Because the added electrons don't come between other electrons and the nucleus, causing interference, you may think that the size of the atoms would remain basically constant. However, as you go from one element to the next, more protons are added to the nucleus, increasing the nuclear charge. The shielding effect of the core electrons is essentially constant as you move from left to right within a period, so the effective nuclear charge increases slightly with each move. This increased effective nuclear charge pulls the atom's electrons, especially the valence electrons, slightly closer to the nucleus, causing the atomic radii (size) of the atoms to slightly decrease when moving from left to right in a period.

Tracing tendencies of ionic radii

Atoms are neutral, but they may gain or lose electrons to form *ions* (atoms or groups of atoms that have an overall charge). To keep things simple, in this chapter I discuss only *monatomic ions,* ions that have only one atom.

If an atom loses electrons, the chemical species that is left has more protons (more positive charges) than electrons (negative charges) and therefore has an overall positive charge. An ion with a positive charge is a *cation.* (I would have preferred the name *dogion,* but they didn't ask my opinion.) A cation has one positive charge for each electron lost.

A cation's radius size (its *ionic radii*) is smaller than a un-ionized atom. The loss of electrons leads to a decrease in the size when compared to the neutral atom because, at least for a representative element, the ion loses an entire energy level. Thus, a potassium ion, K^+, is smaller than a potassium atom. The more electrons removed, the greater the decrease in radius becomes. Therefore, you observe the trend in radii of $Fe > Fe^{2+} > Fe^{3+}$.

When an atom gains electrons, it produces an *anion,* which has a larger ionic radius than the original atom. You observe this trend because the effective nuclear charge (determined by the number of protons) is the same in both the atom and the anion but is spread over an increasing number of electrons. The more electrons added, the greater the increase in size. For oxygen and a couple of its anions, you see the following trend in radii: $O < O^- < O^{2-}$.

For a set of isoelectronic anions (anions with the same number of electrons), such as the anions of nitrogen, oxygen, and fluorine, the number of protons in the nucleus increases as you move across the period from N to O to F. This increase means that the effective nuclear charge also increases, pulling the electrons closer to the nucleus. Therefore, the trend in ionic radii is $F^- < O^{2-} < N^{3-}$.

Eyeing Trends in Ionization Energies

The ease of forming a cation, an ion with a positive charge, varies with the position of the element on the periodic table. To understand the trend in ability to form cations, you must use the concept of ionization energies of the elements. The *ionization energy* is the energy required to remove an electron from a gaseous atom in its ground state. This definition is demonstrated by the thermochemical equation for the formation of a cation from some atom (A) in a gaseous state *(g)*:

$$A(g) \rightarrow A^+(g) + e^- \qquad \Delta H = \text{ionization energy (IE)}$$

In this equation, e^- is the electron being lost and ΔH is the enthalpy (energy) needed to accomplish the removal of the electron.

Removing an electron from an atom always takes energy, so the ionization energy is always *endothermic* (requiring energy). The removal of a second electron requires even more energy (the second ionization energy), the removal of a third electron requires the third ionization energy, and so forth.

The following sections describe some of the trends in ionization energies that you may encounter.

Noting an increase in sequential energy

The amount of energy required for sequential ionization energies (removing more than one electron from an atom) increases with the number of electrons removed. You encounter this in the formation of cations of a +2 or +3 charge, corresponding to the loss of 2 and 3 electrons, respectively. This increase is due to the attractive force between the increasing positive charge of the cation and the electron that is in the process of being removed. Unfortunately, this increase isn't simply linear.

The general relationship between the value of an element's ionization energy and the position of the element on the periodic table is essentially the reverse of the atomic radii (which I discuss in the earlier section "Checking Out the Importance of Size"). Small atoms have high ionization energies, and large atoms have low ionization energies. The size increase of an atom near the bottom of a column on the periodic table means that its outer electrons are further from the nucleus. The greater the distance between the positively charged protons and the negatively charged electrons, the weaker the attraction of the nucleus for the electrons (something physicists call the *inverse square law*). A weaker attraction holding electrons in an energy level means that you can use less energy to overcome the attractive force, which results in a lower ionization energy for the electron. For example, much more energy is required to remove an electron from lithium than from cesium (both alkali metals), because the electron being removed from lithium is much closer to the nucleus and therefore is held more strongly.

Within a period (a row on the periodic table), the ionization energy increases from left to right because the effective nuclear charge is increasing and therefore a greater attractive force must be overcome. However, this trend in ionization energy is more complicated than a simple increase in the effective nuclear charge.

The trends in ionization energy suggest that elements on the left side of the periodic table, especially the lower left, form cations (lose electrons) more readily than the elements on the right. The elements with lower ionization energies are the metals, and the elements with higher ionization energies are the nonmetals. And as I indicate in the chapter on ionic bonding (Chapter 13), metals do lose electrons to form cations.

For example, if you examine the elements from lithium, Li, to neon, Ne, you can see the expected increase in ionization energies from left to right on the periodic table. However, the trend isn't linear; peaks occur at beryllium (Be) and nitrogen (N). Why are the ionization energy values for beryllium and nitrogen higher than expected? If you examine the electron configuration of beryllium, you can see that it is $1s^2 2s^2$ (see Chapter 4 for a discussion of electron configurations). The valence shell of beryllium has a filled 2s subshell. The filling of a subshell leads to additional stability and therefore requires more energy to pull the electron away.

In the case of nitrogen, the 2p sublevel is half-filled. The electrons are in different orbitals and are as widely separated as possible for electrons in the same sublevel. This wide separation minimizes repulsion among the negative charges and stabilizes the half-filled sublevel arrangement of electrons in nitrogen. This kind of stability also leads to an increase in the ionization energy.

Taking stability into consideration

The stability of certain electron configurations affects the general trend in the ionization energies based on size and effective nuclear charge. As I note in the preceding section, stability increases anytime a sublevel is either half-filled or filled.

You can write electronic configurations of cations. For example, a sodium atom loses a single electron to form a Na^+ cation. The electron configuration of Na^+ is

Sodium cation (Na^+) $1s^2 2s^2 2p^6$

Because of quantum mechanical considerations coupled with the effective nuclear charge, the electrons are always lost from the level with the highest principal quantum number. For the representative elements, the last electron to enter is the first electron lost. In the case of the transition metals, the first electron lost is not the last electron to enter. If you use iron, Fe, for example, you can examine the electron configuration of the element and the configurations of the two common iron cations, Fe^{2+} and Fe^{3+}.

Fe $1s^2 2s^2 2p^6 3s^2 3p^6 4s^2 3d^6$

Fe^{2+} $1s^2 2s^2 2p^6 3s^2 3p^6 3d^6$

Fe^{3+} $1s^2 2s^2 2p^6 3s^2 3p^6 3d^5$

The formation of the iron(II) cation, Fe^{2+}, involves the loss of the two 4s electrons because these two electrons have a higher principal quantum number than the last electrons to enter (the 3d electrons). This factor explains why many of the transition metals have a stable 2^+ ion. The iron 3d electrons are not affected until after the removal of the 4s electrons. And this removal deletes energy level 4, making the ions more stable because they have a lower overall energy.

Considering a few exceptions to the rule

For any element, ionization energy increases as an increasing number of electrons are removed. However, some elements do react differently in certain levels. The second ionization energy of sodium, the third ionization energy of magnesium, and the fourth ionization energy of aluminum are significantly higher than you would expect based on their proceeding ionization energies. Apparently something else is going on. If you examine the electron configurations of these three elements, you find

Na $1s^2 2s^2 2p^6 3s^1$

Mg $1s^2 2s^2 2p^6 3s^2$

Al $1s^2 2s^2 2p^6 3s^2 3p^1$

The first ionization energy of sodium, the first two ionization energies of magnesium, and the first three ionization energies of aluminum leave you with the following electron configurations:

Na⁺ $1s^22s^22p^6$

Mg^{2+} $1s^22s^22p^6$

Al^{3+} $1s^22s^22p^6$

For these three elements, removal of an *additional* electron involves removal from the electron configuration $1s^22s^22p^6$, which contains only core electrons (the valence electrons having been already removed in the formation of the cation). This factor is at the root of all three high ionization energy values for the preceding stable ions: The removal of a core electron requires significantly more energy than the removal of valence electrons. In normal chemical reactions, sufficient energy is available to remove valence electrons but not core electrons. Thus, the simple removal of the valence electrons forms common cations of the representative elements.

Considering Trends in Electron Affinities

Although ionization energies describe the formation of cations, chemists use a different term to describe the formation of anions. *Electron affinity* is the energy change accompanying the addition of an electron to a gaseous atom in its ground state. Following is the thermochemical equation for this process, where A is an atom, *(g)* represents the gaseous state, e⁻ is an electron, and ΔH represents the enthalpy (energy) needed to accomplish the addition of the electron.

$$A(g) + e^- \rightarrow A^-(g) \qquad \Delta H = \text{electron affinity (EA)}$$

Unlike the ionization energies, which are always *endothermic* (requiring energy), the first electron affinity of an atom may be endothermic, *exothermic* (producing energy), or even zero. Adding additional electrons, though, is always endothermic. The second and higher electron affinities require energy because, after adding the first electron, the ion is negatively charged and subsequent electrons must overcome the repulsion. Electron affinity values are more difficult to measure than ionization energies; many times they're calculated from thermochemical data.

The general trend of electron affinities is about the same as the trend for ionization energies (discussed in the previous section "Eyeing Trends in Ionization Energies"). Small atoms having a vacancy in the valence shell teamed with a high effective nuclear charge have high electron affinities. Therefore, electron affinities tend to increase going up a column (increasing effective nuclear charge) and moving from left to right in a period (which also increases the effective nuclear charge).

However, electron configurations complicate this trend even more than they complicate the trend for ionization energies. The noble gases have ionization energies, but because of their filled valence shells, they don't have electron affinities. In addition, the alkaline earth metals with their s^2 configurations (no vacancy in the lowest shell of the valence shell) have essentially no electron affinity. Additional complications appear when you consider elements other than the representative elements. For example, the electron affinity of gold is higher than every element other than the halogens. The electron affinity of fluorine is lower than expected because the fluoride ion is too small to accommodate the additional negative charge without a significant amount of repulsion. Because fluorine has a lower than expected electron affinity, the highest electron affinity of all the elements belongs to chlorine. Large positive electron affinities indicated that the negative ion is very stable.

As for the representative elements, the nonmetals have high electron affinities and tend to form anions. The metals have low electron affinities and tend not to form anions.

You can write electron configuration and orbital diagrams for anions, just as you can do so for atoms and cations. The extra electrons are added according to the aufbau principle, which means that they enter the lowest possible orbital (see Chapter 4).

As an example, use oxygen. You find that the atom has the following electron configuration:

O $1s^2 2s^2 2p^4$

When an oxygen atom gains one electron (corresponding to the electron affinity), you get this configuration:

O⁻ $1s^2 2s^2 2p^5$

The addition of a second electron (corresponding to the second electron affinity) gives you the following:

O²⁻ $1s^2 2s^2 2p^6$

The oxide ion, O^{2-}, is now *isoelectronic* (having the same number of electrons) with the noble gas neon, Ne, and like the noble gases, it will not accommodate any additional electrons. It has a filled valence shell, and this situation is extremely stable. The representative elements will only gain electrons until they fill their valence shells (attain an electron arrangement isoelectronic with the next noble gas).

Chapter 17

Examining the Link between Intermolecular Forces and Condensed States

*I*n this chapter, I tell you about *intermolecular forces* (the forces between atoms and/or ions and/or molecules) and *condensed states* (by which I mean liquids and solids). You may think these two topics seem unrelated, but in fact they are related. The state of matter in which a substance is found is dependent on the kinetic energy of the particles of the substance and the strength of the forces between those particles, the intermolecular forces.

Here I show you the different types of intermolecular forces and then discuss the properties of liquids and solids. (I'm not going to talk much about gases; they're pretty well covered in Chapter 6.) So off you go, and may the (intermolecular) forces be with you.

Understanding Types of Intermolecular Forces

Intermolecular forces are those attractive or repulsive forces (interactions) that take place between atoms, molecules, and ions. They are all related to charge, whether a full charge in the case of ions or a partial charge in the case of atoms or molecules. *Remember:* Unlike charges attract; like charges repel.

Being able to recognize whether or not a molecule is polar is important in order to determine which intermolecular forces are important in a particular situation. *Polar* molecules have a partial positive and a partial negative end and thus are dipoles. You may want to glance over Chapters 14 and 15 on covalent bonding and molecular geometry.

Many of the properties of chemical substances, such as solubility, are related to intermolecular forces. Being able to recognize the intermolecular forces present will many times allow you to predict some physical properties and even reactivities.

The following intermolecular forces are listed in order of decreasing strength of interaction.

Bringing ions and dipoles together

Sometimes an ion is attracted to a molecule that is a dipole. If the ion is a *cation* (positive charge), it's attracted to the negative end of the dipole, and if the ion is an *anion* (negative charge), it's attracted to the positive end of the dipole. This interaction occurs quite commonly in aqueous solution in which an ion attracts water molecules.

Suppose, for example, you dissolve some white, crystalline $AlCl_3$ in water. The Al^{3+} cation attracts water molecules because it is a small ion with a large amount of positive charge. (Chemists say it has a large charge density.) Because of its high charge density, it attracts the partial negative end of the water molecule, the oxygen. In fact, it attracts a total of six water molecules, creating the hydrated aluminum ion, $Al(H_2O)_6^{3+}$.

The chloride ions attract the partial positive ends of water molecules, the hydrogens. However, because chloride ions are large ions with only a single negative charge, their charge density is low and so the attractive force is much weaker.

If you carefully evaporate the solution to dryness, you can recover $Al(H_2O)_6Cl_3$, the hexahydrated aluminum chloride. This substance is a hydrate; it has six waters of hydration incorporated into its crystalline structure. Careful heating of most hydrates results in the waters being driven out of the crystalline structure, leaving the anhydrous form.

Mutually attracting dipoles

Dipole-to-dipole attraction is important when dipoles are present. The positive end of one dipole is attracted to the negative end of another. For example, in the extremely reactive chlorine monofluoride gas, the chorine has a partial positive charge, and the more electronegative fluorine has a partial negative charge, allowing a dipole-dipole intermolecular attraction between two of the molecules:

Cl–F — Cl–F

δ+ δ– — δ+ δ–

Dipole-dipole forces tend to be especially important in polar liquids and are considered a strong intermolecular force, but not as strong as ion-dipole.

Drawing close to hydrogen

Hydrogen bonding is a subset of dipole-dipole interaction, but one in which a hydrogen is bonded to an extremely electronegative element (N, O, F). The covalent bond between the hydrogen and these three elements is extremely polar. The hydrogen on one molecule can interact with the O, N, or F on another molecule. This intermolecular force is much stronger than other dipole-dipole forces and so therefore is given its own special name. The strength of this interaction explains a number of unusual properties of substances such as water. As I discuss in Chapter 14, the reason that water has such relatively high boiling and melting points, among other properties, is related to its hydrogen bonding.

In order for an intermolecular force to be hydrogen bonding, a hydrogen atom must be bonded to an N, O, or F, and this hydrogen must interact with an O, N, or F on another molecule.

Uniting through the cloud

Induced dipole intermolecular forces occur when the charge on an ion or a dipole distorts the electron cloud of a nonpolar molecule. A cation attracts the electron cloud, whereas an anion repels it. Either force induces a temporary dipole in the nonpolar molecule. These interactions are fairly weak and tend to occur in solution.

Bonding temporarily with London (dispersion) forces

This intermolecular attraction occurs in all substances but is usually only significant for nonpolar substances. It is created from the momentary distortion of the electron cloud in which the electron density flows to one side of the atom/molecule. That electron cloud is not fixed; I like to compare it to a ball of cotton candy, easily pushed around (but not as tasty).

This distortion causes a very weak temporary dipole. This weak dipole induces a dipole in another molecule. These weak dipoles lead to an attraction. The more electrons present, the larger the cloud and the greater the London force. This interaction is extremely weak and an individual one does not last very long. However, it is strong enough to allow chemists to liquefy nonpolar gases such as hydrogen, H_2, which would be impossible if intermolecular forces didn't attract these molecules.

Grasping the Properties of Liquids

A liquid is a phase in which the randomly orientated particles are in contact. The particles may clump together to exhibit short-range areas of order, but they usually don't last very long. The random orientation allows the liquid to change shape to match that of the container, and because the particles are in contact, a liquid isn't very compressible. In this section, I show you a few important macroscopic properties of liquids. The strength of the intermolecular forces is the key to these different properties.

Resisting an increase: Surface tension

Within the body of a liquid, intermolecular forces pull the molecules in all directions. However, at the liquid's surface, the molecules are pulled down into the body of the liquid and from the sides. As a result, neither molecules nor an attractive force above the surface pulls in that direction. The effect of these unbalanced attractive forces is that the liquid tries to minimize its surface area. The *surface tension* is the resistance of a liquid to an increase in its surface area.

The minimum surface area for a given quantity of matter is a sphere. You may have seen video or pictures of a liquid, usually water, being released in zero gravity. The droplets form little spheres. In a large pool of liquid, where sphere formation is not possible, the surface behaves as if it had a thin, stretched elastic membrane or skin over it. Surface tension requires force to break these attractive forces at the surface. The greater the intermolecular force, the greater the surface tension. Polar liquids, especially those that utilize hydrogen bonding, have a much higher surface tension than nonpolar liquids.

Surface tension is what allows a bug to walk over water. It also allows you to add more water to a glass than its volume. Try it. Carefully add water to a glass and see the water dome at the top. Now touch a toothpick that has been dipped into dishwashing liquid to that dome and see it break. That happens because the dishwashing liquid is a *surfactant,* which disrupts surface tension.

Resisting to flow: Viscosity

Viscosity is the resistance to flow. Suppose you have a glass of paint thinner (nonpolar molecules) and a glass of molasses (polar molecules). Try pouring each into another container. The molasses pours much more slowly than the paint thinner. The two important factors influencing the viscosity of a liquid are as follows:

✔ **Intermolecular forces:** The stronger the intermolecular force, the greater the viscosity. This is the factor in the paint thinner and molasses situation. The mixture of polar molecules that make up molasses attract each other, and the result is a resistance to flow. The size of the molecules also affects intermolecular forces (and molasses). Large and complex molecules have difficulty moving past one another, so the viscosity is high.

✔ **Temperature:** Heat and cold also affect viscosity. If you heat the glass of molasses, it pours more easily because you reduce its viscosity by increasing the kinetic energy of the particles. The higher kinetic energy overcomes the intermolecular attractive forces, causing a lower viscosity. Putting the molasses in the refrigerator causes the opposite effect. The kinetic energy is reduced and the viscosity is increased; hence the phrase, "Slower than molasses in wintertime."

Some liquids have very high viscosities. If the viscosity is high enough, the liquid may not appear to flow at all and may be mistaken for a solid. A high viscosity liquid that appears to be a solid is commonly referred to as an *amorphous* solid. Sometimes scientists refer to these amorphous solids as *glasses* because glass is the most common example. Rubber and charcoal are other examples of amorphous solids.

Climbing the walls: Capillary action

Another property of liquids that's related to intermolecular forces is capillary action. *Capillary action* is the rising of a liquid through a narrow tube against the force of gravity. It's a result of the competition of intermolecular forces within the liquid and the attractive forces between the liquid and the wall of the tube. The stronger the attraction between the liquid and the wall, the higher the level rises.

Mercury has a weak attraction to the walls of a glass tube and so has low capillary action. Water has a strong attraction to the walls of a glass tube and therefore has high capillary action. This capillary action explains why you observe a meniscus with water contained in a thin tube. A *meniscus* is a concave water surface due to the attraction of the water molecules adjacent to the glass walls. Because of mercury's weak attraction to the glass walls, no meniscus is present. However, if you replace the glass tube with a plastic one, water behaves much more like the mercury did in the glass tube because very little attraction exists between the polar water molecules and the nonpolar plastic.

Only the liquid near the walls of the tube is attracted to the walls. The particles farther away from the walls pull the other molecules back. The narrower the tube, the fewer central molecules and the higher the liquid raises in the tube. This property is one of the ways in which water reaches the top of a tall tree.

Warming up: Heat capacity

Heat capacity is the amount of energy needed to cause the temperature of a substance to rise 1K. (See Chapter 10 for a more complete discussion of heat capacity.) The stronger the intermolecular forces between the molecules of a liquid, the more energy that's required to break the forces and the greater the heat capacity. The same reason explains why liquids that have strong intermolecular forces have higher boiling points and vapor pressures than those that do not.

Working with Solids

At the macroscopic level, a *solid* is a substance that has both a definite shape and volume. At the microscopic level, solids are structures in which the particles are vey close to each other. The intermolecular forces have overcome the kinetic energy of the particles, which exhibit no real motion, just vibration.

Solids may be amorphous or crystalline:

✔ **Amorphous:** *Amorphous solids* lack extensive ordering of the particles, so the structure has a lack of regularity. Small regions of order may be separated by large areas of disordered particles. Amorphous solids have no distinct melting point; they simply become softer and softer as the temperature rises.

As noted in "Resisting to flow: Viscosity" earlier in this chapter, liquids with high viscosity that appear to be solid are referred to as amorphous solids. Glass, rubber, and charcoal are examples.

✔ **Crystalline:** *Crystalline solids* display a very regular ordering of the particles in a three-dimensional structure called a *crystal lattice*. In this crystal lattice are repeating units called *unit cells*. Think of a brick wall that is very long, high, and deep. The wall is like the crystal lattice, and each individual brick is like a unit cell.

You can find several types of unit cells in solids. One of the most common is the cubic unit cell, in which all angles are 90 degrees. However, particles can be arranged in this cubic cell in three ways:

- ✔ **Simple unit:** Also called the *primitive* unit cell, the particles are located at the corners of a simple cube.

- ✔ **Body-centered unit:** The particles are located at the corners and another particle is in the center of the cube.

- ✔ **Face-centered unit:** The particles are in the corners of the cell and in the center of each face but not in the center of the unit cell.

The crystalline solids can be further divided into a number of different types based on the type of bonding or intermolecular force that holds the individual particles in the crystal lattice. Scientists know of five types of crystalline solids:

- ✔ **Atomic solids:** In these solids, the particles are held together in a crystal lattice by London forces. The noble gases are the only known examples of atomic solids.

- ✔ **Molecular solids:** In these solids, the particles in their crystal lattices are held in place by London, dipole-dipole, and hydrogen-bonding intermolecular forces. Water and solid methane are examples of molecular solids.

- ✔ **Ionic solids:** In this type, the particles in the crystal lattice are held together by the attraction of cations and anions. This type of crystal lattice tends to be very strong with high melting points, due to the strong attractive forces between the oppositely charged ions. Sodium chloride (NaCl) is an example of an ionic solid. (Have you ever tried to melt salt?)

- ✔ **Metallic solids:** Metal atoms are in a crystal lattice held together by *metallic bonding,* a type of bonding in which the electrons of the metal atoms are delocalized and donated to a "sea of electrons." The electrons are free to move throughout the entire solid, which explains why metals are such good conductors of both electricity and heat. This metallic bonding is also responsible for the luster of metals.

- ✔ **Network (covalent) solids:** These solids have their atoms in the crystal lattice held together by covalent bonds. They tend to have a very extensive structure, and, in fact, you can think of a network solid as one large molecule. Diamonds are examples of network solids. That Hope diamond is a single large molecule! Graphite and silicon dioxide are also examples of network solids.

Deciphering Phase Diagrams

A *phase diagram* is a graph representing the relationship of all the states of matter of a specific substance. The most common type of phase diagram relates the states to temperature and pressure. The pressure is the vertical axis and the temperature is the horizontal axis. Different substances have different ranges of temperature and/or pressure. The lines may be longer or shorter, and the angles may vary, but the solid (S), liquid (L), and gas (G), regions always have the same basic relationship to each other. The phase diagram allows you to predict which state of matter exists at a certain temperature and pressure combination. Figure 17-1 shows a general form of a phase diagram.

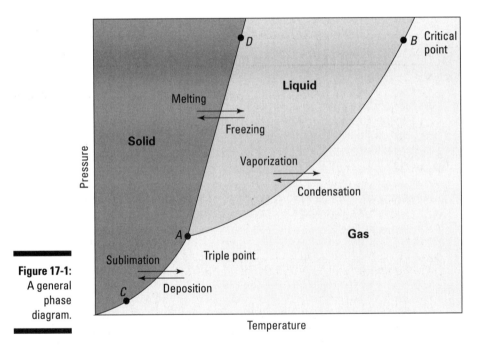

Figure 17-1:
A general
phase
diagram.

You should be able to sketch a simple phase diagram, but knowing how to read the phase diagram for a particular substance is really useful. With a phase diagram you can predict the type of phase change a substance may undergo if the temperature is changed at constant pressure and vice versa.

Notice that the phase diagram has the following three general areas corresponding to the three states of matter (solid, liquid, and gas). The lines separating each area correspond to phase changes:

- **A to C:** This line represents the change in vapor pressure of the solid with temperature for *sublimation* (going directly from a solid to a gas without first becoming a liquid). Crossing the solid-gas line from left to right is sublimation, and the reverse process is deposition. The heat of sublimation or deposition is applicable for changes in this region.

- **A to B:** This line represents the relationship of the vapor pressure of a liquid with pressure. Crossing the liquid-gas line in the diagram from left to right represents vaporization and from right to left represents condensation. Crossing this line involves the heat of vaporization or the heat of condensation.

- **A to D:** This line represents the relationship of the melting point with pressure. Crossing the solid-liquid line in the diagram from left to right is melting (fusion) and from right to left is freezing (solidification). The heats of fusion or solidification are related to these types of changes.

The B point shown in this phase diagram is the *critical point* of the substance, the point beyond which the liquid and gas phases are indistinguishable from each other. Both liquids and gases contain randomly arranged particles; the real difference between liquids and gases is that in a liquid, the particles are in contact, while in a gas the particles are widely separated. Increasing the pressure on a gas pushes its particles closer together. As the pressure continues to increase, the particles eventually come into contact and there is no longer any difference between the gas and a liquid. The two phases become identical in appearance.

At or beyond this critical point, no matter how much pressure is applied, the gas cannot be condensed into a liquid. The temperature at the critical point is the *critical temperature,* and the pressure at the critical point is the *critical pressure.*

Point A is the *triple point* of the substance, the combination of temperature and pressure at which all three states of matter can exist. At the triple point, all phases are present. Any type of change is possible, because melting, boiling, and sublimation (and their reverses) occur simultaneously.

The incredible strength of water

Quite a few activities illustrate the strength of hydrogen bonding in water. One way is to take a clean penny, place it on the table, and carefully add drops of water. You'll be amazed to see how many drops you can add without breaking the surface tension. (This little experiment is also a neat way to check out the magnifying effect of the water dome.)

The other demonstration that I really like that illustrates surface tension is the paperclip experiment. Fill a glass totally full of water and then start adding paperclips slowly. Take a guess how many you can add. In most cases you will greatly underestimate the number of paperclips that you can add to that "full" glass.

Speaking of paperclips, you can float a paperclip on the surface of water if you place it on the surface very carefully. The paperclip is made of steel, which has a density much greater than the density of water, so it should sink, but the hydrogen bonding creates the surface tension that supports the weight of the paperclip. The same phenomenon is also how some insects are able to walk across the surface of a puddle or pond.

Part IV

Environmental Chemistry: Benefits and Problems

IRONICALLY, THE LAST THING PROF. CARUTHERS REMEMBERED WAS EXPLAINING HOW ENERGY IS PRODUCED WHEN MOLECULES COLLIDE.

In this part . . .

Chemistry isn't just something that's done in an academic or industrial lab. Professional chemists aren't the only individuals who do chemistry. *You* do chemistry, too. Chemistry touches your life each and every day.

Chemistry gives all of us great benefits, but it can also give us great problems. Our modern society is complex. Chemistry holds the promise of solving many of the problems facing society, making our lives easier and more meaningful.

In the chapters of this part, I show you some problems that have been created by irresponsible use of chemistry, but also how chemistry has solved many more problems. I discuss air pollution and water pollution — what causes them, what cures them. Finally, I end up discussing nuclear chemistry — the problems and the promises. After reading this chapter, you may virtually glow with excitement.

Chapter 18

Cough! Cough! Hack! Hack! Air Pollution

*T*his chapter looks at the global problem of air pollution. (Personally, I consider the perfume department of a large department store at Christmas time the ultimate in air pollution, but I won't discuss that here.) I show you the chemical problems involved with air pollution, and I explain how air pollution is linked to modern society and its demand for energy and personal transportation.

Seeing Where This Mess Began: Civilization's Effect on the Atmosphere

The air that surrounds the earth — our *atmosphere* — is absolutely necessary for life. The atmosphere provides oxygen (O_2) for respiration and carbon dioxide (CO_2) for *photosynthesis,* the process by which organisms (mainly plants) convert light energy into chemical energy; it moderates the temperature of the earth and plays an active part in many of the cycles that sustain life. The atmosphere is affected by many chemical reactions that take place or exist on earth.

When few humans were on earth, mankind's effect on the atmosphere was negligible. But as the world's population grew, the effect of civilization on the atmosphere became increasingly significant. The Industrial Revolution, which gave rise to the construction of large, concentrated industrial sites, made man's effect on the atmosphere really significant. As humans burned more *fossil fuels* — organic substances such as coal that are found in underground deposits and used for energy — the amount of carbon dioxide (CO_2) and *particulates* (small, solid particles suspended in the air) in the atmosphere increased significantly. Following the Industrial Revolution, humans also began to use more items that released chemical pollutants into the atmosphere, including hairsprays and air conditioners.

The increase in CO_2 and particulates, combined with the increase in pollutants, has disrupted delicate balances in the atmosphere. High concentrations of these atmospheric pollutants have led to a multitude of problems such as *acid deposition,* acidic rain that damages living things, buildings, and statues, and *photochemical smog,* the brown, irritating haze that often sits over Los Angeles and other cities.

Taking a Closer Look at the Earth's Atmosphere

The earth's atmosphere is divided into several layers: the troposphere, the stratosphere, the mesosphere, and the thermosphere. I want to focus on the two layers closest to the earth — the troposphere and stratosphere — because they're the layers affected the most by humans. They're also the layers that have the greatest direct effect on human life.

The troposphere: What humans affect the most

The *troposphere* lies next to the earth and contains the gases you breathe and depend on for survival. The troposphere is composed of about 78.1 percent nitrogen (N_2), 20.9 percent oxygen (O_2), 0.9 percent argon (Ar), 0.03 percent carbon dioxide (CO_2), and smaller amounts of various other gases. The troposphere also contains varying amounts of water vapor. These gases are held tight to the earth by the force of gravity. If a balloonist were to rise high into the troposphere, she would find the atmospheric gases much thinner due to the decreased pull of gravity on the gases. This effect indicates that the dense layer of gases held tight to the earth is more at risk from the effects of pollution.

The troposphere is the layer where weather occurs. It's also the layer that takes the brunt of both natural and man-made pollution because of its proximity to the earth.

Nature pollutes the atmosphere to a certain extent — with noxious hydrogen sulfide (H_2S) and particulate matter from volcanoes, and the release of organic compounds from plants such as pine trees. But these pollutants have a minor effect on the troposphere. Humankind, on the other hand, pollutes the troposphere with a large amount of chemicals from automobiles, power plants, and industries. Acid rain and photochemical smog are some of the results of man-made pollutants.

The stratosphere: Protecting humans with the ozone layer

Above the troposphere is the *stratosphere,* which is where jets and high-altitude balloons fly. The atmosphere is much thinner in this layer because of the decreasing pull of gravity. Few of the heavier pollutants are able to make it to the stratosphere, because gravity holds them tight and close to the surface of the earth. The protective *ozone layer* resides in the stratosphere; this protective barrier absorbs a large amount of harmful ultraviolet (UV) radiation from the sun and keeps it from reaching the earth.

Even though heavier pollutants don't make their way to the stratosphere, this layer isn't immune to the effects of mankind. Some lighter manmade gases do make it into the stratosphere, where they attack the protective ozone layer and destroy it. This destruction can have far-reaching effects on humans because UV radiation is a major cause of skin cancer.

A chemical substance can be both a good guy and a bad guy. The only difference is where and in what concentration it's found. For example, a person can overdose on water if he drinks enough of it. The same goes with the ozone in the stratosphere. On one hand, it shields us from harmful UV radiation. But on the other, it can be an irritant and destroy rubber products (see "Breathing Brown Air: Photochemical Smog" for details).

Getting the Lowdown on the Ozone Layer

The ozone layer absorbs almost 99 percent of the ultraviolet radiation that reaches the earth from the sun. It protects us from the effects of too much ultraviolet radiation, including sunburns, skin cancer, cataracts, and premature aging of the skin. Because of the ozone layer, most of us can enjoy the outdoors without head-to-toe protection.

How is ozone (O_3) formed? Well, oxygen in the *mesosphere* — the part of the earth's atmosphere between the stratosphere and the *thermosphere* (the layer that extends to outer space) — is broken apart by ultraviolet radiation into highly reactive oxygen atoms. These oxygen atoms combine with oxygen molecules in the stratosphere to form ozone.

$O_2(g)$ + ultraviolet radiation $\rightarrow 2\ O(g)$

$O_2(g) + O(g) \rightarrow O_3(g)$

Unfortunately, the ozone layer isn't a perfect defense. The following sections explain how the ozone layer is compromised.

Explaining how the ozone reacts to gases

As a society, humans release many gaseous chemicals into the atmosphere. Many of the gaseous chemicals rapidly decompose through reactions with each other, or they react with the water vapor in the atmosphere to form compounds such as acids that fall to earth in the rain (see "'I'm Meltinggggggg!' — Acid Rain," later in this chapter). Besides forming acid rain, some of these chemicals also form photochemical smog (see "Breathing Brown Air: Photochemical Smog," later in this chapter).

But these reactions occur rather quickly, and people can deal with them in a variety of ways, many of which are related to breaking the series of reactions that produce the pollutant by stopping the release of a critical chemical into the air.

Some classes of gaseous chemical compounds are rather *inert* (inactive and unreactive), so they remain within the atmosphere and cause negative effects for quite a while. One such troublesome class of compounds is the *chlorofluorocarbons,* gaseous compounds composed of chlorine, fluorine, and carbon. These compounds are commonly called *CFCs.*

Because CFCs are relatively unreactive, they were extensively used in the past as refrigerants for such items as refrigerators and automobile air conditioners (Freon-12), foaming agents for plastics such as Styrofoam, and propellants for the aerosol cans of such consumer goods as hair spray and deodorants. As a result, they were released into the atmosphere in great quantities. Over the years, the CFCs have diffused into the stratosphere, and they're now doing damage to it.

Seeing how CFCs hurt the ozone layer

Although CFCs don't react much when they're close to earth — they're pretty inert — most scientists believe that they react with the ozone in the atmosphere and then harm the ozone layer in the stratosphere.

The reaction occurs in the following way:

1. **A typical chlorofluorocarbon, CF_2Cl_2, reacts with ultraviolet radiation, and a highly reactive chlorine atom is formed.**

   $$CF_2Cl_2(g) + UV \text{ light} \rightarrow CF_2Cl(g) + Cl(g)$$

2. **The reactive chlorine atom reacts with ozone in the stratosphere to produce oxygen gas molecules and chlorine oxide (ClO).**

   $$Cl(g) + O_3(g) \rightarrow O_2(g) + ClO(g)$$

   This reaction destroys the ozone layer. If things stopped here, the problems would actually be minimal.

3. **The chlorine oxide (ClO) can then react with another oxygen atom in the stratosphere to produce an oxygen molecule and a chlorine atom; the newly created oxygen molecule and chlorine atom are now available to start the whole ozone-destroying process all over again.**

   $$ClO(g) + O(g) \rightarrow O_2(g) + Cl(g)$$

So one CFC molecule can initiate a process that can destroy many molecules of ozone.

Are harmful CFCs still produced?

The problem of ozone depletion was identified in the 1970s. As a result, the governments of many industrialized nations began to require the reduction of the amount of CFCs and halons released into the atmosphere. (Halons, which contain bromine in addition to fluorine and chlorine, were commonly used as fire-extinguishing agents, especially in fire extinguishers used around computers.)

CFCs were banned for use as propellants in aerosol cans in many countries, and the CFCs used in the production of plastics and foams were recovered instead of released into the air. Laws were enacted to ensure that the CFCs and halons used as refrigerants were recovered during the recharging and repair of units. In 1991, DuPont started producing refrigerants that weren't harmful to the ozone layer. And in 1996, the United States, along with 140 other countries, stopped producing chlorofluorocarbons altogether.

Unfortunately, though, these compounds are extremely stable. They'll remain in our atmosphere for many years. If the damage man has done to the ozone layer isn't too great, it may replenish itself (like new skin grows to replace sunburned skin). But it may well be years before the ozone layer returns to its former composition.

Comprehending the Greenhouse Effect

When most people think about air pollutants, they think of such chemicals as carbon monoxide, chlorofluorocarbons, or hydrocarbons. Yet carbon dioxide, the product of animal respiration and the compound used by plants in the process of photosynthesis, can also be considered a pollutant if present in abnormally high amounts.

In the late 1970s and early 1980s, scientists realized that the average temperature of the earth was increasing. They determined that an increase in carbon dioxide (CO_2) and a few other gases, such as chlorofluorocarbons (CFCs), methane (CH_4, a hydrocarbon), and water vapor (H_2O), were responsible for the slight increase in temperature through a process called the *greenhouse effect* (so named because the gases serve pretty much the same purpose as the glass walls and roof of a greenhouse: keeping heat in). The gases themselves are called *greenhouse gases*.

Here's how the greenhouse effect works: As radiation from the sun travels through the earth's atmosphere, it strikes the earth, heating the land and water. Some of this solar energy is sent back (reflected) into the atmosphere as heat (infrared radiation), which is then absorbed by certain gases (CO_2, CH_4, H_2O, and CFCs) in the atmosphere. These gases, in turn, warm the atmosphere. This process helps to keep the temperature of the earth and atmosphere moderate and relatively constant, and as a result, we don't experience dramatic day-to-day temperature fluctuations. So, in general, the greenhouse effect is a good thing, not a bad thing.

But if an excess of carbon dioxide and other greenhouse gases are floating around, too much heat gets trapped in the atmosphere. The atmosphere heats up, leading to the disruption of many of the delicate cycles of the earth. This process is commonly called *global warming* or *global climate change*, and it's currently happening with the earth's atmosphere.

Modern civilization depends on burning fossil fuels (coal, natural gas, or petroleum) for energy. People burn coal and natural gas to produce electricity, gasoline to run internal combustion engines, and natural gas, oil, wood, and coal to heat homes. In addition, industrial processes burn fuel to produce heat. As a result of all this burning of fossil fuels, the carbon dioxide level in the atmosphere has risen from 318 parts per million (ppm) in 1960 to 388 ppm in 2010. (For a discussion of the concentration unit *ppm*, see Chapter 9.) The excess carbon dioxide has led to an increase of about a half-degree in the average temperature of the atmosphere.

A half-degree increase in the average temperature of the atmosphere may not sound like much, but this global warming trend may have serious effects on several of the ecological systems of the world:

✔ The rising atmospheric temperature may melt ice masses and cause sea levels around the world to rise. Rising sea levels may result in the loss of coastal land (Houston may sink into the Gulf of Mexico) and make many more people vulnerable to *storm surges* (extremely damaging rushes of seawater that occur during very bad storms).

✔ The increased temperature may affect the growth patterns of plants.

✔ The tropical regions of the world may increase and lead to the spread of tropical diseases.

Breathing Brown Air: Photochemical Smog

Smog is a generic word people use to describe the combination of smoke and fog that's often irritating to breathe. The two major types of smog are

✔ London smog
✔ Photochemical smog

London smog

London smog is a gaseous atmospheric mixture of fog, soot, ash, sulfuric acid (H_2SO_4 — battery acid), and sulfur dioxide (SO_2). The name comes from the air pollution that plagued London in the early part of the 20th century. The burning of coal for heat in the highly populated city caused this smog. The dangerous mixture of gases and soot from the coal stoves and furnaces killed more than 8,000 people in London in 1952.

Electrostatic precipitators and scrubbers (see "Charge them up and drop them out: Electrostatic precipitators" and "Washing water: Scrubbers," later in this chapter), combined with filters, have been effective in reducing the release of soot, ash, and sulfur dioxide into the atmosphere and have reduced the occurrence of London smog.

Photochemical smog

Photochemical smog is produced after sunlight initiates certain chemical reactions involving unburned hydrocarbons and oxides of nitrogen (commonly shown as NO_x — which stands for a mixture of NO and NO_2). The common automobile engine produces both of these compounds when it's running.

Photochemical smog is the brown haze that sometimes makes it difficult to see in such cities as Los Angeles, Salt Lake City, Denver, and Phoenix. (This smog is sometimes called *Los Angeles smog* — sometimes sunny California isn't so sunny.) These cities are especially vulnerable to photochemical smog; they have a large number of automobiles, which emit the chemicals that react to produce the smog, and they're surrounded by mountain ranges. The mountain ranges and the westward winds create an ideal condition for thermal inversions, which trap the pollutants close to the cities. (In a *thermal inversion,* a layer of warmer air moves in over a layer of cooler air. The warm air traps the cooler air and its pollutants close to the ground. The process can be compared to sheets trapping certain noxious gases in a bed. The gaseous pollutants are trapped and can't move higher in the atmosphere. They stay close to us humans, causing all kinds of problems.)

The chemistry of photochemical smog is still not crystal clear (pun intended), but scientists do know the basics that go into creating the smog. Nitrogen from the atmosphere is oxidized to nitric oxide in internal combustion engines and then released into the atmosphere through the engines' exhaust systems:

$$N_2(g) + O_2(g) \rightarrow 2\,NO(g)$$

The nitric oxide is oxidized to nitrogen dioxide by atmospheric oxygen:

$$2\,NO(g) + O_2(g) \rightarrow 2\,NO_2(g)$$

Nitrogen dioxide, a brownish gas, is irritating to the eyes and lungs. It absorbs sunlight and then produces nitric oxide and highly reactive oxygen atoms:

$$NO_2(g) + \text{sunlight} \rightarrow NO(g) + O(g)$$

These reactive oxygen atoms quickly react with diatomic (two-atom) oxygen gas molecules in the air to produce ozone (O_3):

$$O(g) + O_2(g) \rightarrow O_3(g)$$

This ozone is what acts as a shield against ultraviolet radiation in the stratosphere. But when it's down closer to the earth, it acts as a powerful irritant to the eyes and lungs. It attacks rubber, causing it to harden, and thus shortens the life of automobile tires and weather stripping. It also affects crops such as tomatoes and tobacco.

The unburned hydrocarbons from auto exhaust also react with the oxygen atoms and ozone to produce a variety of organic aldehydes that are also irritants. These hydrocarbons can react with diatomic oxygen and nitrogen dioxide to produce peroxyacetylnitrates (PANs):

$$\text{Hydrocarbons}(g) + O(g) + NO_2(g) \rightarrow \text{PANs}$$

These PANs are also eye and lung irritants; they tend to be very reactive, causing damage to living organisms. The combination of the brown nitrogen dioxide, the ozone, and the PANs is photochemical smog. It reduces visibility and is a major cause of respiratory problems. And, unfortunately, controlling it has been difficult.

Auto emissions have been closely monitored, and strict controls have been put into place to minimize the amount of unburned hydrocarbons released into the atmosphere. The Clean Air Act of 1990 was passed to help reduce hydrocarbon emissions from automobiles. The catalytic converter was developed to help react the unburned hydrocarbon and produce a less-dangerous emission of carbon dioxide and water. (As a side benefit, lead had to be eliminated from gasoline because it "poisoned" the catalyst and made the catalytic converter useless. The big campaign to "get the lead out" removed a major source of the deadly heavy metal from the environment.)

Although such measures as catalytic converters and activated carbon canisters, which are used to help reduce gasoline fumes, have been somewhat effective, photochemical smog still presents a problem. Until mankind develops an acceptable substitute for the internal combustion engine or requires mass transit, photochemical smog will remain with us for years to come.

"I'm Meltingggggg!" — Acid Rain

The Wicked Witch in *The Wizard of Oz* dissolved in water. Sometimes buildings do the same because of the action of acid rain on the limestone and marble. Rainwater is naturally acidic (with a pH less than 7) as a result of the dissolving of carbon dioxide in the moisture of the atmosphere and the forming of carbonic acid. (See Chapter 11 for more about carbonic acid as well as the pH scale.) This interaction results in rainwater having a pH of around 5.6. The term *acid rain,* or *acid deposition,* is used to describe a situation in which rainfall has a much lower (more acidic) pH than can be explained by the simple dissolving of carbon dioxide. The following gives you an overview of acid rain including what comprises it and how to clean it.

Don't drink the water: What's in acid rain

Acid rain is formed when certain pollutants in the atmosphere, primarily oxides of nitrogen and sulfur, dissolve in the moisture of the atmosphere and fall to earth as rain with a low pH value.

Oxides of nitrogen (NO, NO_2, and so on) are produced naturally during lightning discharges in the atmosphere. This is one way that nature "fixes" nitrogen, or puts it in a form that can be used by plants. However, man adds tremendously to the local amount of atmospheric nitrogen oxides through the use of automobiles. The internal combustion engine reacts the gasoline hydrocarbons with the oxygen in the air, producing carbon dioxide (and carbon monoxide) and water. But the nitrogen that's present in the air (about 78 percent of the air is nitrogen) may also react with the oxygen at the high temperatures present in the engine. This reaction can produce nitric oxide (NO), which is then released into the atmosphere:

$$N_2(g) + O_2(g) \rightarrow 2 NO(g)$$

As the NO enters the atmosphere, it reacts with additional oxygen gas to produce nitrogen dioxide (NO$_2$):

$$2 NO(g) + O_2(g) \rightarrow 2 NO_2(g)$$

This nitrogen dioxide can then react with the water vapor in the atmosphere to form nitric and nitrous acids:

$$2 NO_2(g) + H_2O(g) \rightarrow HNO_3(aq) + HNO_2(aq)$$

These dilute acid solutions fall to earth as rain with a low pH value, generally in the 4.0 to 4.5 range (although rains with a pH as low as 1.5 have been reported).

A significant amount of acid rain in the eastern part of the United States is caused by oxides of nitrogen, but the acid rain of the Midwest and West is caused by mostly oxides of sulfur, which are primarily generated by power plants and the burning of coal and oil. Sulfur-containing compounds are found as impurities in coal and oil, sometimes as high as 4 percent by weight. These compounds, when burned, produce a sulfur dioxide gas (SO$_2$). Many millions of tons of SO$_2$ are released into the atmosphere each year from power-generating plants. The SO$_2$ reacts with the water vapor in the atmosphere to produce sulfurous acid (H$_2$SO$_3$) and with the oxygen in the atmosphere to produce sulfur trioxide (SO$_3$):

$$SO_2(g) + H_2O(g) \rightarrow H_2SO_3(aq)$$
$$2 SO_2(g) + O_2(g) \rightarrow 2 SO_3(g)$$

This sulfur trioxide then reacts with the moisture in the atmosphere to produce sulfuric acid (H$_2$SO$_4$), which is the same acid found in your car battery:

$$SO_3(g) + H_2O(g) \rightarrow H_2SO_4(aq)$$

So the sulfurous and sulfuric acids that are dissolved in the rainwater form the acid rain that falls to the earth. Anyone for a bath in battery acid?

The acids formed in the atmosphere can travel many hundreds of miles before falling to earth as acid rain and leaving their mark on both nonliving and living things. The acids of the rain react with the iron in buildings and automobiles, causing them to corrode. The acids also destroy the details of fine works of art when they react with marble statues and limestone buildings to form soluble compounds that wash away. (Want to see this in action? Put a drop of vinegar, an acid, on a piece of marble and then watch the bubbles form as the acid dissolves the marble. Careful, though. Don't try this on anything too valuable — maybe use that ugly marble cheese slicer that Aunt Gertrude gave you last Christmas.)

You probably won't be surprised to hear that acid rain has a bad effect on vegetation. Acid rain has been identified as the major cause of death of many trees and even whole forests. Even if trees aren't killed immediately by acid rain, forests sometimes grow slower because of its effects. The growth may be hindered by the release of aluminum from the soil, which interferes with the absorption of nutrients, or it may be slowed by the bacteria found in the soil.

In addition, acid rain has altered the ecosystems of many lakes in Canada and the United States. Fish kills have been reported, and entire species of fish have vanished from certain lakes. In fact, the ecosystems of entire lakes have been destroyed by acid rain, rendering the lakes lifeless.

Steps have been taken to reduce acid rain and its effects. Increasing fuel efficiency and the use of pollution control devices on automobiles have helped reduce the amount of nitrogen oxide released into the atmosphere. But fossil fuel power plants produce the most tonnage of acid-causing pollutants. A number of controls have been adopted to decrease the amount of sulfur-containing gases released into the atmosphere, including electrostatic precipitators and scrubbers, which are discussed in the following two sections. But although they've been effective in reducing the amount of acid-causing material released into the atmosphere, much more still needs to be done before the problem of acid rain is reduced to a manageable level.

Charge them up and drop them out: Electrostatic precipitators

When you were a child, did you ever run a comb through your hair on a cold winter morning and then use it to pick up little scraps of paper? An electrostatic precipitator does much the same thing.

Electrostatic precipitators give a negative electrical charge to pollutant particles. The sides of the precipitator have a positive charge, so the negative particles are then pulled to the positively charged walls. They stick to the walls and accumulate there. Then they can be removed (it's like sweeping out those dust bunnies from under the bed).

In one type of electrostatic precipitation system, the SO_2 produced by the burning of fossil fuels is reacted with lime (CaO) to produce solid calcium sulfite ($CaSO_3$):

$$SO_2(g) + CaO(s) \rightarrow CaSO_3(s)$$

The finely divided calcium sulfite is electrostatically precipitated and collected. It can then be disposed of properly in a chemical landfill.

Washing water: Scrubbers

Scrubbers are thingies that remove impurities from pollutant gases by using a fine spray of water to trap the gases as an aqueous solution or force them through a reacting mixture. The process is similar to using a water spray to settle dust in arid regions.

You can use a scrubber as an especially efficient system for removing sulfur dioxide by forcing the SO_2 through a slurry of magnesium hydroxide and converting it to magnesium sulfite, which can then easily be collected:

$$SO_2(g) + Mg(OH)_2(aq) \rightarrow MgSO_3(s) + H_2O(l)$$

Is the quality of air getting better?

The quality of air in cities such as Los Angeles has improved over the last 20 years. Pollution controls have reduced the oxides of nitrogen and unburned hydrocarbons released by automobiles, and the levels of photochemical smog have been significantly reduced.

Pollution controls have also reduced the levels of sulfur dioxides released from power plants, which has helped lower the occurrence of acid rain. In addition, the ban on CFCs released into the atmosphere should eventually have an effect on ozone depletion. So in many respects, yes, the quality of our air is improving.

But humans are still releasing tremendous amounts of carbon dioxide into the atmosphere.

And we're simultaneously using up large amounts of the earth's valuable plant and animal material (biomass), which tends to consume this excess carbon dioxide.

The effect on the environment is debated on a daily basis. All scientists agree that the effect is negative. The question is simply a matter of degree. If mankind can reduce its dependence on fossil fuels for electricity and heat by using solar, nuclear, or perhaps even fusion power, then maybe we'll be able to make advances in reducing the amount of carbon dioxide released into the atmosphere. This strategy, combined with limits on the destruction of forests, may bring the problem of global warming under control.

Chapter 19

Examining the Ins and Outs of Water Pollution

. . . .

. . . .

Water is absolutely necessary to every human's survival. After all, the human body is about 70 percent water. Most of the water on earth, however, is found as seawater. Only about 2 percent of the water on the earth is fresh water, and a little more than three-quarters of that is in the form of ice and glaciers. But it's that very small amount of fresh water suitable for drinking (*potable* water) that most people are concerned about.

I'm sure you're quite aware of the water you drink and the water you use for bathing, cooking, and watering your lawn. But unless you live in a rural agricultural area, you may not think much about the water used to grow the plants and animals for food.

In addition, water is used to carry waste products from homes, to generate electricity, and in chemical reactions and cooling towers. And then there's recreation — boating, swimming, and fishing. All these things depend on an adequate supply of good, pure water.

But where does water come from? How does it get contaminated, and how does it get cleaned? These questions are really chemistry questions. In fact, most pollution, whether it be air pollution or water pollution, can be related to chemistry, and I discuss the relationship between chemistry and water pollution in this chapter. So sit back, grab your glass of water, and dive in.

Where Does Water Come from, and Where Is It Going?

The actual amount of water on earth is relatively constant, but its location and purity may vary. Water moves throughout the environment by what is called the water cycle, or the hydrologic cycle. Figure 19-1 diagrams this cycle. The following sections outline this cycle in greater depth.

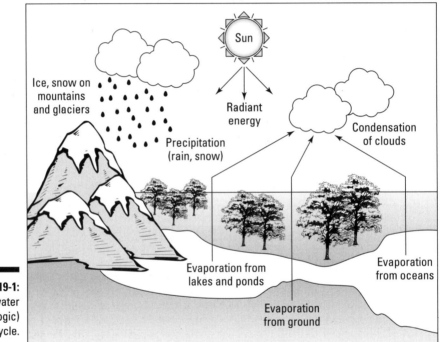

Figure 19-1:
The water
(hydrologic)
cycle.

Evaporate, condense, repeat

Water *evaporates* (goes from a liquid to a gas when heated) from lakes, streams, oceans, trees, and even humans. As water evaporates, it leaves behind any contaminants that it may have accumulated. (The salt on your sweatband and cap is left behind by evaporating sweat.) This process of evaporation is one of nature's ways of purifying water.

The water vapor may then travel many miles, or it may stay relatively local, depending on the prevailing winds. Sooner or later, the vapor condenses (goes from a gas to a liquid when cooled) and falls back to the earth as rain, snow, or sleet.

Following the water

Water may fall to earth and collect in a lake or stream. If it does, it eventually finds its way back to the sea. If it falls onto the land, it can form *runoff* and eventually enter a lake or stream, or it can soak into the ground and become *groundwater*. The porous layer of soil and rock that holds the groundwater forms a zone called an *aquifer*. This zone provides us with a good source of groundwater. Wells tap into these aquifers.

Human activities can affect this water cycle. Cutting vegetation can increase the rate of runoff, causing less water to become absorbed into the soil. Man-made dams and reservoirs increase the surface area available for water evaporation. Using more groundwater than can be replenished may deplete the aquifers and lead to water shortages. And society can contaminate the water in a wide variety of ways that I discuss in this chapter.

Taking a Closer Look at Water: A Most Unusual Substance

Water is a polar molecule. Chapter 14 covers polar molecules in detail, but here's a quickie overview relating to water: The oxygen in water (H_2O) has a higher *electronegativity* (attraction for a bonding pair of electrons) than the hydrogen atoms, so the bonding electrons are pulled in closer to the oxygen. The oxygen end of the water molecule then acquires a partial negative charge, and the hydrogen atoms take on a partial positive charge. When the partially positively charged hydrogen (where's my editor on *that* clunker of a phrase?) of one water molecule is attracted to the partially negatively charged oxygen of another water molecule, a rather strong interaction between the water molecules can occur. This interaction is called a hydrogen bond (H-bond). This is not to be confused with a hydrogen *bomb,* a very different thing. Figure 19-2 shows the hydrogen bonds that occur in water.

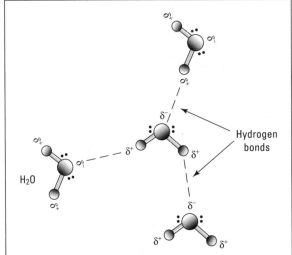

Figure 19-2:
Hydrogen
bonding in
water.

Hydrogen bonds, caused by the polar covalent bonds of water molecules, give water some very unusual properties:

- ✔ **Water has a very high surface tension.** The water molecules at the surface of the water are only attracted downward into the body of the liquid. The molecules in the body of the liquid, on the other hand, are attracted in all different directions. Bugs and small lizards can walk across water because they don't exert enough force to break the surface tension. The high surface tension of water also means that evaporation rates are really less than you'd expect.

- ✔ **Water becomes a liquid at temperatures commonly found on earth.** The boiling point of a liquid is normally related to its molecular weight. Substances that have molecular weights close to the molecular weight of water (18 g/mol) boil at far lower temperatures; these substances become gases at normal room temperature.

- ✔ **Ice, the solid state of water, floats when placed in water.** Normally, solids have a higher density than their corresponding liquids, because the particles are closer together in solids. When water freezes, however, it's locked into a crystal lattice that has large holes incorporated into it by its hydrogen bonds. So the density of ice is less than that of water (see Figure 19-3).

  The floating property of ice is one of the reasons that life, in all its diversity and magnitude, is able to exist on earth. If ice were denser than water in the winter, the water at the top of lakes would freeze and sink. Then more water would freeze and sink, and so on. Pretty soon, the lake would be frozen solid, destroying most of the life — such as plants and fish — in the lake. Instead, ice floats and forms an insulating layer over the water, which allows life to exist, even in the winter.

Figure 19-3:
The
structure
of ice.

✔ **Water has a relatively high heat capacity.** The *heat capacity* of a substance is the amount of heat a substance can absorb or release in order to change its temperature 1 degree Celsius. Water's heat capacity is almost 10 times greater than iron and 5 times greater than aluminum. This means that lakes and oceans can absorb and release large amounts of heat without a dramatic change in temperature, which moderates the temperature on earth. Lakes absorb heat during the day and release it at night. Without water's high heat capacity, the earth would undergo dramatic swings in temperature during its day/night cycle.

✔ **Water has a high heat of vaporization.** The *heat of vaporization* of a liquid is the amount of energy needed to convert a gram of the liquid to a gas. Water has a heat of vaporization of 54 calories per gram (see Chapter 10 for more about the calorie, a metric unit of heat). This high heat of vaporization allows us to rid our bodies of a great deal of heat when sweat is evaporated from the skin. This property also helps to keep the climate on earth relatively moderate without extreme short-term swings.

✔ **Water is an excellent solvent for a large number of substances.** In fact, water is sometimes called the universal solvent, because it dissolves so many things. Water is a polar molecule, so it acts as a solvent for polar solutes. It dissolves ionic substances easily; the negative ends of the water molecules surround the cations (positively charged ions), while the positive ends of the water molecules surround the anions (negatively charged ions). (Turn to Chapter 13 for specifics on ions, cations, and anions.) With the same process, water can dissolve many polar covalent compounds, such as alcohols and sugars (see Chapter 14 for more on these types of compounds). This is a desirable property, but it also means that water dissolves many substances that are not desirable to us or that make water unusable. Chemists lump all those substances together under the terms *pollutants* or *contaminants*.

Identifying Some Common Water Pollutants

Because water is such an excellent solvent, it easily picks up unwanted substances from a variety of sources. Figure 19-4 shows some sources of water contamination.

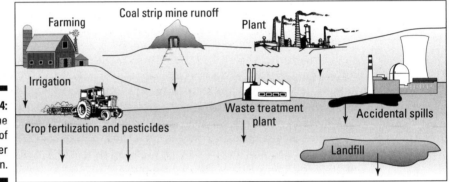

Figure 19-4: Some sources of water pollution.

I call Figure 19-4 Pollution Place, because it shows so many pollution sources in the same place. Naturally, you won't find this many pollution sites this close together in too many places in the United States.

Pollution sources are normally classified as point sources or non-point sources:

- ✔ *Point sources* are pollution sites that have a definite identifiable source. Discharges from a chemical industry or raw sewage from wastewater treatment plants are common examples of point-source pollution. Point sources are easy to identify, control, and regulate. The Environmental Protection Agency (EPA) is the governmental agency in the United States that regulates point sources.

- ✔ *Non-point sources* are pollution sources that are rather diffuse in nature. Good examples of this type of pollution are water contamination caused by agricultural runoff or acid rain. Controlling and regulating this type of pollution is much more difficult because you can't identify a particular company or individual as the polluter. In recent years, federal and state agencies have attempted to address non-point source pollution. The Clean Water Action Plan of 1998 was one such attempt that focused on watersheds and runoffs.

The following sections look at some of the more common pollutants that can be found in water.

We really didn't get the lead out: Heavy metal contamination

Water supplies are closely monitored for heavy metals, because they tend to be very toxic. Major sources of heavy metal contamination include landfills, industries, agriculture, mining, and old water distribution systems. Following are some metals that cause metal contamination:

- ✔ **Lead:** This heavy metal pollutant has received a lot of press in the past. Large amounts of lead entered the environment from the use of leaded gasoline: The tetraethyl lead that was used to boost the gasoline's octane was oxidized in the combustion process, and a large amount of lead was emitted through exhaust systems. Rain runoff carried the lead into streams where it was deposited. Another source of lead was old pipes in municipal buildings and homes. These pipes were commonly joined with a lead solder that then leached lead into the drinking water.

- ✔ **Mercury:** This element is released into the aquatic environment from mercury compounds used to treat seeds from fungus and rot. Runoff from fields washes the mercury compounds into the surface water and sometimes into the groundwater supply.

- ✔ **Chromium:** The automobile is also an indirect source of this heavy metal contaminant. Chromium compounds (such as CrO_4^{2-}) are used in chrome plating for bumpers and grills. This plating also requires the use of the cyanide ion (CN^-), another major pollutant. These contaminants used to be discharged directly into streams, but now they're either pretreated to reduce to a less-toxic form or precipitated (formed into a solid) and disposed of in landfills.

- ✔ **Metals from mining:** Mining also adds to the heavy metal pollution problem. As the earth is mined, deposits of minerals, which contain metals, are exposed. If the chemicals used in extracting ore or coal deposits are acidic, then the metals in the minerals are dissolved, and they may make their way into the surface water and sometimes the groundwater. This problem is sometimes controlled with a process that isolates mine drainage and then treats it to remove the metal ions.

Biological concentration is a problem that occurs when industries release heavy metal ions (from the preceding pollutants and others) into the waterways. As metal ions move through the ecosystem, they become more and more concentrated. (The same thing happens with radioisotopes — see Chapter 20 for details.) The ions may be released at a very low concentration level, but by the time they move up the food chain to us, the concentration may be at the toxic level. This situation happened in Minamata Bay, Japan. An industry was dumping mercury metal into the bay. As the metal moved through the ecosystem, it was eventually converted to the extremely toxic methylmercury compound. People died as a result of the toxins, and others became permanently affected.

Raining down acid

Oxides of nitrogen and sulfur can combine with the moisture in the atmosphere to form rain that can be highly acidic — acid rain. This rain can affect the pH of lakes and streams and has been known to seriously affect aquatic life. In fact, it's made some lakes devoid of life altogether.

Acid rain is a good example of a non-point source of pollution. Pinpointing a single entity as the cause is difficult. Air pollution controls have decreased the amount of acid rain produced, but it's still a major problem. (If you want more info on acid rain, flip to Chapter 18.)

Getting sick off infectious agents

This category of contamination includes fecal coliform bacteria from human wastes and the wastes of birds and other animals. Fecal coliform bacteria was previously a major problem in the United States and most parts of the world. Resulting epidemics of typhoid, cholera, and dysentery were common. Treatment of wastewater has minimized this problem in industrialized nations, but it's still a definite problem in underdeveloped nations.

Many experts think that more than three-quarters of the sicknesses in the world are related to biological water contaminants. And even now in the United States, beaches and lakes are still closed at times because of biological contamination.

Stricter controls on municipal water treatment, septic tanks, and runoff from feedlots can help decrease the biological contamination of our water.

Leaking landfills and LUST

Landfills — both the public and hazardous chemical kinds — are a major source of groundwater contamination. The landfills constructed today require special liners to prevent hazardous materials from leaching into the groundwater. Monitoring equipment is also required to confirm that the hazardous materials don't leak from the landfills. However, very few landfills in the United States have liners and monitoring systems.

Many landfills contain *VOCs* (volatile organic chemicals). This group of chemicals includes benzene and toluene (both carcinogens), chlorinated hydrocarbons, such as carbon tetrachloride, and trichloroethylene, which previously was used as a dry-cleaning solvent. Even though these compounds are not very soluble in water, they do accumulate at the parts-per-million level. Their long-term effect on human health is unknown at this time.

Most people think of toxic wastes in terms of an industrial dump, but the municipal landfill is becoming a more popular site for the disposal of hazardous household wastes. Every year, tons of the following toxic materials are placed in commercial landfills:

- Batteries containing heavy metals like mercury
- Oil-based paints containing organic solvents
- Motor oil containing metals and organic compounds
- Gasoline containing organic solvents
- Automobile batteries containing sulfuric acid and lead
- Antifreeze containing organic solvents
- Household insecticides containing organic solvents and pesticides
- Fire and smoke detectors containing radioactive isotopes
- Nail-polish remover containing organic solvents

Some cities and states are trying to reduce the amount of toxic substances released into the environment by providing special collection sites for materials such as used motor oil. But much more needs to be done.

LUST (leaking underground storage tanks) is another source of VOCs. The major culprit? Old, rusted gasoline storage tanks — especially from filling stations that have long been out of business. Less than a gallon of leaked gasoline can contaminate the water supply of a midsized town. Recent federal regulations have required the identification and replacement of leaking tanks, but abandoned service stations have become a major problem. As many as 200,000 tanks still need to be replaced.

The problem of hazardous materials in landfills and the contamination of the water supply prompted Congress to pass the Superfund program. This program was designed to identify and clean up potentially harmful landfills and dumps. Some progress has been made, but thousands of dump sites may need to be cleaned at a monumental cost to the taxpayer.

The alternatives to landfills are recycling and incineration. Some of the material that commonly goes into a public landfill can be recycled — paper, glass, aluminum, and some plastics, for example — but more needs to be done. Incineration of some materials can be accomplished with the generation of electricity. Modern incineration produces very little air pollution.

Seeping pollution from farms

Many types of water pollution are associated with the agricultural industry. For example, the excessive use of fertilizers, which contain nitrate and phosphate compounds, have caused a dramatic increase in the growth of algae and plants in lakes and streams. This increased growth may interfere with the normal cycles that occur in these aquatic systems, causing them to age prematurely — a process known as *eutrophication*.

In addition, pesticides used in treating crops may be released into the waterways. These pesticides, especially the organo-phosphorus ones, may undergo biological concentration (see "We really didn't get the lead out: Heavy metal contamination," earlier in this chapter). Because of the effects of DDT, the United States has banned its use, but it's still manufactured and sold overseas.

The release of soil and silt into the waterways is another form of pollution associated with the agricultural industry. The soil builds up in the water and interferes with the normal cycles of lakes and streams. It also carries agricultural chemicals into the waterways.

Polluting with heat: Thermal pollution

People usually think of things like lead, mercury, toxic organic compounds, and bacteria as being major pollutants. However, heat can also be a major pollutant. The solubility of a gas in a liquid *decreases* as the temperature increases (see Chapter 6 for more about the solubility of gases), so warm water doesn't contain as much dissolved oxygen as cool water. And how is this related to pollution? The amount of oxygen in water has a direct impact on aquatic life. The reduction of the dissolved oxygen content of water caused by heat is called *thermal pollution*.

Industries, especially those that generate electric power, use a tremendous amount of water to cool steam and condense it back to water. This water is normally taken from a lake or stream, used in the cooling process, and then returned to the same body of water. If the heated water is returned directly to the lake or stream, the increase in temperature may cause the oxygen levels to decrease below those required for the survival of certain types of fish. The increased temperature may also trigger or repress natural cycles of aquatic life, such as spawning.

Federal regulations prohibit the release of heated water back into lakes or streams. Industries cool the water by allowing it to remain in pools or running it over the outside of cooling towers. The cooling towers help the water release its heat to the atmosphere. Both of these methods, however, lose a lot of water to evaporation. (And believe me, some places in the United States certainly don't need the increased humidity.)

Using up oxygen: BOD

If organic material (such as raw sewage, organic chemicals, or a dead cow) finds its way into the water, it decays. The decaying process is basically the oxidation of the organic compounds by *aerobic bacteria,* or oxygen-consuming bacteria, into simpler molecules such as carbon dioxide and water.

The process requires dissolved oxygen (DO) from the water. The amount of oxygen needed to oxidize the organic material is called the *biological oxygen demand* (BOD), and it's normally measured in parts per million (ppm) of oxygen needed. If the BOD is too high, too much dissolved oxygen is used, and not enough oxygen remains for the fish. Fish kills occur, leading to an even higher BOD.

In extreme cases, not enough oxygen is left for the aerobic bacteria to survive, so another group of bacteria, *anaerobic bacteria,* assumes the job of decomposing the organic material. Anaerobic bacteria doesn't use oxygen in the water; instead, it uses oxygen that's in the organic compounds. Anaerobic bacteria reduces the waste instead of oxidizing it. (See Chapter 7 for a discussion of oxidation and reduction.) The bad news is that anaerobic bacteria decomposes organic matter into foul-smelling compounds such as hydrogen sulfide (H_2S), ammonia, and amines.

In order to stop overloading the BOD of the waterways, most chemical industries pretreat (normally with oxidization) their waste chemicals before releasing them into the water. Cities and towns do the same with their wastewater treatment plants.

Getting the Stink out of Wastewater

The days when towns and cities in the United States could dump raw, untreated sewage into the waterways are largely over. Every once in a while, a treatment plant malfunctions or becomes overloaded due to some natural disaster and has to dump raw sewage, but those situations are few and far between.

In much of the world (South Asia and most of Africa, for example), very little sewage gets treated. But in the United States, sewage gets at least primary treatment; it often gets secondary and tertiary treatment, too. Figure 19-5 diagrams both the primary and secondary treatment of sewage.

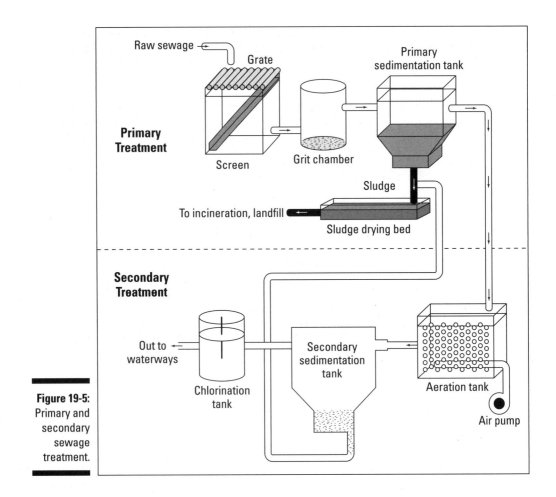

The following sections give you the lowdown on how water treatment plants work to remove bacteria and other particles in sewage so people can use the water.

Primary sewage treatment

In *primary sewage treatment,* raw sewage basically undergoes settling and filtration. The sewage first goes through a grate and screen system to remove large items. (I don't even want to talk about what those items are.) It then moves through a grit chamber where more material is filtered. Finally, it goes to a primary sedimentation tank, where the material is treated with solutions of aluminum sulfate and calcium hydroxide. The two solutions form aluminum hydroxide, a gelatinous precipitate (solid) that accumulates dirt and bacteria as it settles. This primary treatment removes about 50 to 75 percent of the solids, but it only reduces the biological oxygen demand (BOD) by about 30 percent.

If primary treatment is all that the wastewater undergoes, then sometimes chlorine is added to kill a majority of the bacteria before the wastewater is returned to the waterways. It still contains a high BOD, though. So if the waterway is a lake or slow-moving stream, then the high BOD causes problems — especially if a number of towns use the same type of sewage treatment. The problems can be prevented with secondary sewage treatment.

Secondary sewage treatment

In *secondary sewage treatment*, bacteria and other microorganisms are given the opportunity to decompose the organic compounds in the wastewater. Because aerobic bacteria (oxygen-consuming bacteria) produces products that are less noxious than those produced by anaerobic bacteria (bacteria that uses oxygen in the organic compounds instead of oxygen in the water), the sewage is commonly aerated in order to provide the needed oxygen.

Both the primary and secondary processes produce a material called *sludge*, which is a mixture of particulate matter and living and dead microorganisms. The sludge is dried and disposed of by incineration or in a landfill. It can even be spread as a fertilizer on certain types of cropland.

But even secondary treatment can't remove some substances that are potentially harmful to the environment. These substances include certain organic compounds, certain metals such as aluminum, and fertilizers such as phosphates and nitrates. Tertiary sewage treatment can be used to remove these substances.

Tertiary sewage treatment

Tertiary sewage treatment is essentially a chemical treatment that removes the fine particles, nitrates, and phosphates in wastewater. The basic procedure is adjusted for the specific substance to be removed. Activated charcoal filtration, for example, is used to remove most of the dissolved organic compounds. And alum, $Al_2(SO_4)_3$, is used to precipitate phosphate ions by dissolving and freeing the aluminum cation.

$$Al^{3+}(aq) + PO_4^{3-}(aq) \rightarrow AlPO_4(s)$$

Ion exchange (replacing one ion with another), reverse osmosis (Chapter 9), and distillation (heating a mixture and collection the various gaseous components) are also occasionally used with this type of treatment. All these procedures are relatively expensive, though, so tertiary treatment isn't done unless really necessary.

Even after tertiary treatment is completed, the wastewater must still be disinfected before it's released back into the waterways. It's commonly disinfected by bubbling chlorine gas (Cl_2) into the water. Chlorine gas is an extremely powerful oxidizing agent, and it's very effective at killing the organisms responsible for cholera, dysentery, and typhoid. But the use of chlorine has come under question lately. If residual organic compounds are in the wastewater, they can be converted into chlorinated hydrocarbons. Several of these compounds have been shown to be carcinogenic. The levels of these compounds are being closely monitored during testing of wastewater.

Ozone (O_3) can also be used to disinfect wastewater. It's effective at killing viruses that chlorine can't kill. It's more expensive, however, and doesn't provide the residual protection against bacteria.

Treating Drinking Water

Many people tend to take for granted the availability of good drinking water. Most people of the world aren't as fortunate as people in the United States and Canada, who benefit from good treatment of drinking water.

The water is brought in from a lake, stream, or reservoir and initially filtered to remove sticks, leaves, dead fish, and such. The turbidity (haziness) that's commonly present in river or lake water is removed through treatment with a mixture of alum (aluminum sulfate) and lime (calcium hydroxide), which forms gelatinous aluminum hydroxide and traps the suspended solids. This treatment is basically the same as what's used in wastewater treatment plants (see "Primary sewage treatment").

Then the water is filtered again to remove the solid mass of fine particles (called a *flocculate* or *floc*) left over from the initial filtering treatment. Chlorine is added to kill any bacteria in the water. Then it's run through an activated charcoal filter that absorbs (collects on its surface) and removes substances responsible for taste, odor, and color. Fluoride may be added at this time to help prevent tooth decay. Finally, the purified water is collected in a holding tank, ready for your use.

Chapter 20

Nuclear Chemistry: It'll Blow Your Mind

*M*ost of this book deals, in one way or the other, with chemical reactions. And when I talk about these reactions, I'm really talking about how the valence electrons (the electrons in the outermost energy levels of atoms) are lost, gained, or shared. I mention very little about the nucleus of the atom because, to a very large degree, it's not involved in chemical reactions.

But in this chapter, I do discuss the nucleus and the changes it can undergo. I talk about radioactivity and the different ways an atom can decay. I discuss half-lives and show you why they're important in the storage of nuclear waste products. I also discuss nuclear fission in terms of bombs, power plants, and the hope that nuclear fusion holds for mankind.

Like some of you reading this book, I'm a child of the Atomic Age. I actually remember open-air testing of nuclear weapons. I remember being warned not to eat snow because it might contain fallout. I remember friends building fallout shelters, A-bomb drills at school, and X-ray machines in shoe stores. (I never did order those X-ray glasses, though!) And I remember radioactive Fiesta stoneware and radium watch hands. When I was growing up, atomic energy was new, exciting, and scary. And it still is.

Understanding Basic Atomic Structure: It All Starts with the Atom

To understand nuclear chemistry, you need to know the basics of atomic structure. Chapter 4 drones on and on about atomic structure, if you're interested. This section just provides a quickie brain dump.

The *nucleus,* that dense central core of the atom, contains both protons and neutrons. Electrons are outside the nucleus in energy levels. Protons have a positive charge, neutrons have no charge, and electrons have a negative charge. A neutral atom contains equal numbers of protons and electrons. But the number of neutrons within an atom of a particular element can vary. Atoms of the same element that have differing numbers of neutrons are called *isotopes.* Figure 20-1 shows the symbolization chemists use to represent a specific isotope of an element.

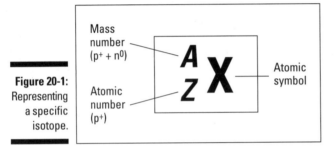

Figure 20-1:
Representing
a specific
isotope.

Mass
number
$(p^+ + n^0)$

Atomic
number
(p^+)

$$^A_Z X$$

Atomic
symbol

In the figure, X represents the symbol of the element found on the periodic table, Z represents the *atomic number* (the number of protons in the nucleus), and A represents the *mass number* (the sum of the protons and neutrons in that particular isotope). If you subtract the atomic number from the mass number $(A - Z)$, you get the number of neutrons in that particular isotope. A short way to show the same information is to simply use the element symbol (X) and the mass number (A) — for example, U-235.

Defining Radioactivity and Man-Made Radioactive Decay

For purposes of this book, I define *radioactivity* as the spontaneous decay of an unstable nucleus. An unstable nucleus may break apart into two or more other particles with the release of some energy (see "Initiating Reactions: Nuclear Fission," later in this chapter, for more info on this process). This breaking apart can occur in a number of ways, depending on the particular atom that's decaying.

You can often predict one of the particles of a radioactive decay by knowing the other particle. Doing so involves something called *balancing the nuclear reaction.* (A *nuclear reaction* is any reaction involving a change in nuclear structure.)

Balancing a nuclear reaction is really a fairly simple process. The first thing you should know is how to represent a reaction:

Reactants → Products

Reactants are the substances you start with, and products are the new substances being formed. The arrow, called a *reaction arrow,* indicates that a reaction has taken place.

For a nuclear reaction to be balanced, the sum of all the atomic numbers on the left-hand side of the reaction arrow must equal the sum of all the atomic numbers on the right-hand side of the arrow. The same is true for the sums of the mass numbers.

Suppose you're a scientist performing a nuclear reaction by bombarding a particular isotope of chlorine (Cl-35) with a neutron. (Work with me here. I'm just trying to get to a point.) You observe that an isotope of hydrogen, H-1, is created along with another isotope, and you want to figure out what the other isotope is.

The equation for this example is

$$^{35}_{17}\text{Cl} + ^{1}_{0}\text{n} \rightarrow ? + ^{1}_{1}\text{H}$$

To figure out the unknown isotope (represented by ?), do the following steps:

1. **Balance the atomic numbers in the equation.**

   The sum of the atomic numbers on the left is 17 (17 + 0), so you want the sum of the atomic numbers on the right to equal 17, too. Right now, you've got an atomic number of 1 on the right; 17 – 1 is 16, so that's the atomic number of the unknown isotope. This atomic number identifies the element as sulfur (S).

2. **Balance the mass numbers in the equation.**

   The sum of the mass numbers on the left is 36 (35 + 1), and you want the sum of the mass numbers on the right to equal 36, too. Right now, you've got a mass number of 1 on the right; 36 – 1 is 35, so that's the mass number of the unknown isotope.

   Now you know that the unknown isotope is a sulfur isotope (S-35). And here's what the balanced nuclear equation looks like:

   $$_{17}^{35}\text{Cl} + _{0}^{1}\text{n} \rightarrow _{16}^{35}\text{S} + _{1}^{1}\text{H}$$

This equation represents a nuclear *transmutation,* the conversion of one element into another. Nuclear transmutation is a process human beings control. S-35 is an isotope of sulfur that doesn't exist in nature. It's a *man-made isotope.* Alchemists, those ancient predecessors of chemists, dreamed of converting one element into another (usually lead into gold), but they were never able to master the process. Chemists are now able, sometimes, to convert one element into another.

Radioactively Decaying the Natural Way

Certain isotopes aren't stable. In an unstable isotope, the nucleus undergoes nuclear decay and breaks apart. Sometimes the product of that nuclear decay is unstable itself and undergoes nuclear decay, too. For example, when U-238 (one of the radioactive isotopes of uranium) initially decays, it produces Th-234, which decays to Pa-234. The decay continues until, finally, after a total of 14 steps, Pb-206 is produced. Pb-206 is stable, and the decay sequence, or series, stops.

Before I show you *how* radioactive isotopes decay, I want to briefly explain *why* a particular isotope decays. The nucleus has all those positively charged protons shoved together in an extremely small volume of space. All those protons are repelling each other. The forces that normally overpower the repelling protons and hold the nucleus together, the "nuclear glue," sometimes can't do the job, and so the nucleus breaks apart, undergoing nuclear decay.

All elements with 84 or more protons are unstable; they eventually undergo decay. Other isotopes with fewer protons in their nucleus are also radioactive. The radioactivity corresponds to the neutron/proton ratio in the atom. If the neutron/proton ratio is too high (caused by too many neutrons or too few protons), the isotope is said to be *neutron rich* and is, therefore, unstable. Likewise, if the neutron/proton ratio is too low (it has too few neutrons or too many protons), the isotope is unstable. The neutron/proton ratio for a certain element must fall within a certain range for the element to be stable, which is why some isotopes of an element are stable and others are radioactive.

Naturally occurring radioactive isotopes decay in three primary ways:

- Alpha particle emission
- Beta particle emission
- Gamma radiation emission

In addition, a couple of less common types of radioactive decay are

- Positron emission
- Electron capture

The following sections examine these ways in more depth.

Alpha particle emission

An *alpha particle* is defined as a positively charged particle of a helium nuclei. I hear ya: *Huh?* Try this: An alpha particle is composed of two protons and two neutrons, so it can be represented as a helium-4 atom. As an alpha particle breaks away from the nucleus of a radioactive atom, it has no electrons, so it has a 2+ charge. Therefore and to-wit, it's a *cation,* a positively charged ion — see Chapter 4.

But electrons are basically free — easy to lose and easy to gain. So, normally, an alpha particle is shown with no charge because it very rapidly picks up two electrons and becomes a neutral helium atom instead of an ion.

Large, heavy elements, such as uranium and thorium, tend to undergo alpha emission. This decay mode relieves the nucleus of two units of positive charge (two protons) and four units of mass (two protons + two neutrons). What a process! Each time an alpha particle is emitted, four units of mass are lost. I wish I could find a diet that would allow me to lose four pounds at a time!

Radon-222 (Rn-222) is another alpha particle emitter, as shown in the following equation:

$$^{222}_{86}\text{Rn} \rightarrow \, ^{218}_{84}\text{Po} + \, ^{4}_{2}\text{He (an alpha particle)}$$

Here, radon-222 undergoes nuclear decay with the release of an alpha particle. The other remaining isotope must have a mass number of 218 (222 – 4) and an atomic number of 84 (86 – 2), which identifies the element as polonium (Po). (If this subtraction stuff confuses you, check out how to balance equations in the section "Defining Radioactivity and Man-Made Radioactive Decay," earlier in this chapter.)

Beta particle emission

A *beta particle* is essentially an electron that's emitted from the nucleus. (Now, I know what you're thinking — electrons aren't in the nucleus. Keep reading to find out how they can be formed in this nuclear reaction.) Iodine-131 (I-131), which is used in the detection and treatment of thyroid cancer, is a beta particle emitter:

$$^{131}_{53}\text{I} \rightarrow \, ^{131}_{54}\text{Xe} + \, ^{0}_{-1}\text{e (a beta particle)}$$

Here, the iodine-131 gives off a beta particle (an electron), leaving an isotope with a mass number of 131 (131 – 0) and an atomic number of 54 (53 – (–1)). An atomic number of 54 identifies the element as Xenon (Xe).

Notice that the mass number doesn't change in going from I-131 to Xe-131, but the atomic number increases by one. In the iodine nucleus, a neutron was converted (decayed) into a proton and an electron, and the electron was emitted from the nucleus as a beta particle.

$$^{1}_{0}\text{n} \rightarrow \, ^{1}_{+1}\text{p} + \, ^{0}_{-1}\text{e}$$

Isotopes with a high neutron/proton ratio often undergo beta emission, because this decay mode allows the number of neutrons to be decreased by one and the number of protons to be increased by one, thus lowering the neutron/proton ratio.

Gamma radiation emission

Alpha and beta particles have the characteristics of matter: They have definite masses, occupy space, and so on. However, because there is no mass change associated with gamma emission, I refer to gamma emission as *gamma radiation emission*. Gamma radiation is similar to X-rays — high energy, short wavelength radiation. Gamma radiation commonly accompanies both alpha and beta emission, but it's usually not shown in a balanced nuclear reaction. Some isotopes, such as cobalt-60 (Co-60), give off large amounts of gamma radiation. Co-60 is used in the radiation treatment of cancer. The medical personnel focus the gamma rays on the tumor, thus destroying it.

Positron emission

Although positron emission doesn't occur with naturally occurring radioactive isotopes, it does occur naturally in a few man-made ones. A *positron* is essentially an electron that has a positive charge instead of a negative charge. A positron is formed when a proton in the nucleus decays into a neutron and a positively charged electron. The positron is then emitted from the nucleus.

$$_{+1}^{1}p \rightarrow {}_{0}^{1}n + {}_{+1}^{0}e$$

This process occurs in a few isotopes, such as potassium-40 (K-40), as shown in the following equation:

$$_{19}^{40}K \rightarrow {}_{18}^{40}Ar + {}_{+1}^{0}e$$

The K-40 emits the positron, leaving an element with a mass number of 40 (40 – 0) and an atomic number of 18 (19 – 1). An isotope of argon (Ar), Ar-40, has been formed.

If you watch *Star Trek,* you may have heard about antimatter. The positron is a tiny bit of antimatter. When it comes in contact with an electron, both particles are destroyed with the release of energy. Luckily, not many positrons are produced. If a lot of them were produced, you'd probably have to spend a lot of time ducking explosions.

Electron capture

Electron capture is a rare type of nuclear decay in which an electron from the innermost energy level (the 1s — see Chapter 4) is captured by the nucleus. This electron combines with a proton to form a neutron. The atomic number decreases by one, but the mass number stays the same.

$$_{+1}^{1}p + {}_{-1}^{0}e \rightarrow {}_{0}^{1}n$$

The following equation shows the electron capture of polonium-204 (Po-204):

$$_{84}^{204}Po + {}_{-1}^{0}e \rightarrow {}_{83}^{204}Bi + X\text{-rays}$$

The electron combines with a proton in the polonium nucleus, creating an isotope of bismuth (Bi-204).

The capture of the 1s electron leaves a vacancy in the 1s orbitals. Electrons drop down to fill the vacancy, releasing energy not in the visible part of the electromagnetic spectrum but in the X-ray portion. This emission of X-rays is what reveals that electron capture is taking place.

Figuring Out When Radioactive Decay Happens: Half-Lives

If you could watch a single atom of a radioactive isotope, U-238, for example, you wouldn't be able to predict when that particular atom might decay. It might take a millisecond, or it might take a century. You simply have no way to tell.

But if you have a large enough sample — what mathematicians call a *statistically significant sample size* — a pattern begins to emerge. It takes a certain amount of time for half the atoms in a sample to decay. It then takes the same amount of time for half the remaining radioactive atoms to decay, and the same amount of time for half of those remaining radioactive atoms to decay, and so on. The amount of time it takes for one-half of a sample to decay is called the *half-life* of the isotope, and it's given the symbol $t_{1/2}$. This process is shown in Table 20-1.

Table 20-1	Half-Life Decay of a Radioactive Isotope
Half-Life	*Percent of Radioactive Isotope Remaining*
0	100.00
1	50.00
2	25.00
3	12.50
4	6.25
5	3.12
6	1.56
7	0.78
8	0.39
9	0.19
10	0.09

The half-life decay of radioactive isotopes is not linear. For example, you can't find the remaining amount of an isotope as 7.5 half-lives by finding the midpoint between 7 and 8 half-lives. This decay is an example of an exponential decay, shown in Figure 20-2.

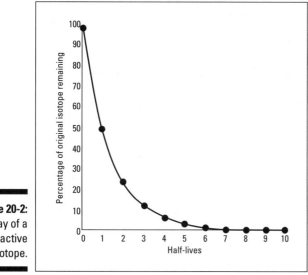

Figure 20-2:
Decay of a
radioactive
isotope.

If you want a firmer grasp of how to calculate a half-life, the importance of half-lives, and how to use half-lives, keep on reading the following sections.

Determining half-lives

Calculations involving simple multiples of half-lives is straightforward. You can use the table in the previous section as a guide. The amounts don't have to be expressed in percentages; you can use grams, micrograms, or a unit of radioactive emission.

If you want to find times or amounts that are not associated with a simple multiple of a half-life, you can use these two equations:

$$(1)\ t_{1/2} = \frac{0.693}{k}$$

$$(2)\ \ln = \left(\frac{N_t}{N_o}\right) = -kt$$

In the equation, *ln* stands for the *natural logarithm* (not the base 10 log; it's that ln button on your calculator, not the log button), N_o is the amount of radioactive isotope that you start with, N_t is the amount of radioisotope left at some time (t), k is a constant, and $t_{1/2}$ is the half-life of the radioisotope. If you know the half-life and the amount of the radioactive isotope that you start with, you can use this equation to calculate the amount remaining radioactive at any time. For example, suppose you find a wooden handle that has 60.0 percent the radioactivity as that of C-14 ($t_{1/2}$ of 5,730 years). You can use the two preceding equations to calculate the age of the handle (when the tree died or was cut down). Because I know the half-life, I can use equation (1) to calculate k:

$$t_{1/2} = \frac{0.693}{k}$$

$$5{,}730y = \frac{0.693}{k}$$

$$k = \frac{1.21 \times 10^{-4}}{y}$$

I can now substitute k into equation (2) and solve for *t:*

$$\ln\left(\frac{N_t}{N_o}\right) = -kt$$

$$\ln\left(\frac{60}{100}\right) = -\frac{1.21 \times 10^{-4}}{y}t$$

$$-0.5108 = -\frac{1.21 \times 10^{-4}}{y}t$$

$$t = 4.2 \times 10^3 y$$

I can also use these same two equations to calculate the half-life if I know the amount (N_t) and the time (t).

Half-lives may be very short or very long. Table 20-2 shows the half-lives of some typical radioactive isotopes.

Table 20-2	Half-Lives of Some Radioactive Isotopes	
Radioisotope	*Radiation Emitted*	*Half-Life*
Kr-94	Beta	1.4 seconds
Rn-222	Alpha	3.8 days
I-131	Beta	8 days
Co-60	Gamma	5.2 years
H-3	Beta	12.3 years
C-14	Beta	5,730 years
U-235	Alpha	4.5 billion years
Re-187	Beta	70 billion years

Safe handling

Knowing about half-lives is important because it enables you to determine when a sample of radioactive material is safe to handle. The rule is that a sample is safe when its radioactivity has dropped below detection limits. And that occurs at 10 half-lives. So if radioactive iodine-131 ($t_{1/2}$ = 8 days) is injected into the body to treat thyroid cancer, it'll be "gone" in 10 half-lives, or 80 days.

This stuff is important to know when using radioactive isotopes as *medical tracers,* which are taken into the body to allow doctors to trace a pathway or find a blockage, or in cancer treatments. They need to be active long enough to treat the condition, but they should also have a short enough half-life so that they don't injure healthy cells and organs.

Radioactive dating

A useful application of half-lives is *radioactive dating.* No, radioactive dating has nothing to do with taking an X-ray tech to the movies. It has to do with figuring out the age of ancient things.

Carbon-14 (C-14), a radioactive isotope of carbon, is produced in the upper atmosphere by cosmic radiation. The primary carbon-containing compound in the atmosphere is carbon dioxide, and a very small amount of carbon dioxide contains C-14. Plants absorb C-14 during photosynthesis, so C-14 is incorporated into the cellular structure of plants. Plants are then eaten by animals, making C-14 a part of the cellular structure of all living things.

As long as an organism is alive, the amount of C-14 in its cellular structure remains constant. But when the organism dies, the amount of C-14 begins to decrease. Scientists know the half-life of C-14 (5,730 years, listed in Table 20-2), so they can figure out how long ago the organism died (refer to the problem in "Determining half-lives").

Radioactive dating using C-14 has been used to determine the age of skeletons found at archeological sites. It was also used to date the Shroud of Turin, a piece of linen in the shape of a burial cloth that contains an image of a man. Many thought that it was the burial cloth of Jesus, but in 1988, radiocarbon dating determined that the cloth dated from around AD 1200–1300. Even though no one knows how the image of the man was imprinted on the Shroud, C-14 dating has proven that it's not the death cloth of Jesus.

Carbon-14 dating can only be used to determine the age of something that was once alive. It can't be used to determine the age of a moon rock or a meteorite. For nonliving substances, scientists use other isotopes, such as potassium-40.

Initiating Reactions: Nuclear Fission

In the 1930s, scientists discovered that some nuclear reactions can be initiated and controlled (see "Defining Radioactivity and Man-Made Radioactive Decay," earlier in this chapter). Scientists usually accomplished this task by bombarding a large isotope with a second, smaller one — commonly a neutron. The collision caused the larger isotope to break apart into two or more elements, which is called *nuclear fission*. The nuclear fission of uranium-235 is shown in the following equation:

$$^{235}_{92}U + ^{1}_{0}n \rightarrow ^{142}_{56}Ba + ^{91}_{36}Kr + 3^{1}_{0}n$$

Reactions of this type also release a lot of energy. Where does the energy come from? Well, if you make *very* accurate measurements of the masses of all the atoms and subatomic particles you start with and all the atoms and subatomic particles you end up with and then compare the two, you find that some mass is "missing." Matter disappears during the nuclear reaction. This loss of matter is called the *mass defect*. The missing matter is converted into energy.

The following sections provide an overview of just what you need to know during your Chem I course about nuclear fission.

Calculating chain reactions and critical mass

You can actually calculate the amount of energy produced during a nuclear reaction with a fairly simple equation developed by Einstein: $E = mc^2$. In this equation, E is the amount of energy produced, m is the "missing" mass, or the mass defect, and c is the speed of light, which is a rather large number. The speed of light is squared, making that part of the equation a *very* large number that, even when multiplied by a small amount of mass, yields a *large* amount of energy. For example, when a mole of U-235 decays to Th-234, the mass defect is 5×10^{-6} kg, which when converted to energy amounts to 5×10^{11} joules!

Notice that one neutron was used, but three were produced. These three neutrons, if they encounter other U-235 atoms, can initiate other fissions, producing even more neutrons. It's the old domino effect. In terms of nuclear chemistry, it's a continuing cascade of nuclear fissions called a *chain reaction*. The chain reaction of U-235 is shown in Figure 20-3.

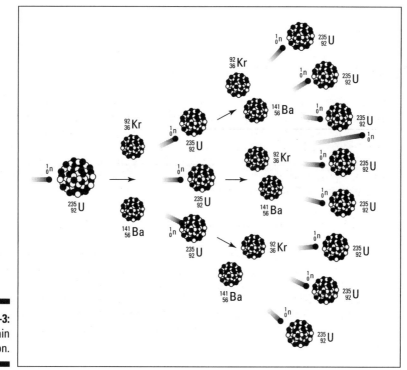

Figure 20-3: Chain reaction.

Atomic bombs (big bangs that aren't theories)

Because of the tremendous amount of energy released in a fission chain reaction, the military implications of nuclear reactions were immediately realized. The first atomic bomb was dropped on Hiroshima, Japan, on August 6, 1945.

In an atomic bomb, two pieces of a fissionable isotope are kept apart. Each piece by itself is subcritical. When time comes for the bomb to explode, conventional explosives force the two pieces together to cause a critical mass. The chain reaction is uncontrolled, releasing a tremendous amount of energy almost instantaneously.

The real trick, however, is to control the chain reaction, releasing its energy slowly so that ends other than destruction can be achieved.

This chain reaction depends on the release of more neutrons than were used during the nuclear reaction. If you were to write the equation for the nuclear fission of U-238, the more abundant isotope of uranium, you'd use one neutron and only get one back out. You can't have a chain reaction with U-238. But isotopes that produce an *excess* of neutrons in their fission support a chain reaction. This type of isotope is said to be *fissionable,* and only two main fissionable isotopes are used during nuclear reactions — uranium-235 and plutonium-239.

A certain minimum amount of fissionable matter is needed to support a self-sustaining chain reaction, and it's related to those neutrons. If the sample is small, then the neutrons are likely to shoot out of the sample before hitting a U-235 nucleus. If they don't hit a U-235 nucleus, no extra electrons and no energy are released. The reaction just fizzles. The minimum amount of fissionable material needed to ensure that a chain reaction occurs is called the *critical mass.* Anything less than this amount is called *subcritical.*

Controlling reactions: Nuclear power plants

The secret to controlling a chain reaction is to control the neutrons. If the neutrons can be controlled, then the energy can be released in a controlled way. That's what scientists have done with nuclear power plants.

In many respects, a nuclear power plant is similar to a conventional fossil fuel power plant. In this type of plant, a fossil fuel (coal, oil, natural gas) is burned, and the heat is used to boil water, which in turn is used to make steam. The steam is then used to turn a turbine that's attached to a generator that produces electricity.

The big difference between a conventional power plant and a nuclear power plant is that the nuclear power plant produces heat through nuclear fission chain reactions.

Comprehending how nuclear power plants make electricity

Most people believe that the concepts behind nuclear power plants are tremendously complex. That's really not the case. Nuclear power plants are very similar to conventional fossil fuel plants.

The fissionable isotope is contained in fuel rods in the reactor core. All the fuel rods together comprise the critical mass. Control rods, commonly made of boron or cadmium, are in the core, and they act like neutron sponges to control the rate of radioactive decay. Operators can stop a chain reaction completely by pushing the control rods all the way into the reactor core, where they absorb all the neutrons. The operators can then pull out the control rods a little at a time to produce the desired amount of heat.

A liquid (water or sometimes liquid sodium) is circulated through the reactor core, and the heat generated by the fission reaction is absorbed. The liquid then flows into a steam generator, where steam is produced as the heat is absorbed by water. This steam is then piped through a steam turbine that's connected to an electric generator. The steam is condensed and recycled through the steam generator. This forms a closed system; that is, no water or steam escapes — it's all recycled.

The liquid that circulates through the reactor core is also part of a closed system. This closed system helps ensure that no contamination of the air or water takes place. But sometimes problems do arise.

Eyeing the downsides to nuclear power plants

The United States has approximately 100 nuclear reactors, which produce a little more than 20 percent of the country's electricity. In France, almost 80 percent of the country's electricity is generated through nuclear fission. Nuclear power plants have certain advantages. No fossil fuels are burned (saving fossil-fuel resources for producing plastics and medicines), and the process uses no combustion products, such as carbon dioxide, sulfur dioxide, and so on, to pollute the air and water. But significant problems are associated with nuclear power plants, and I discuss them in the following sections.

Cost issues: They aren't cheap

Nuclear power plants are expensive to build and operate. The electricity that's generated by nuclear power costs about twice as much as electricity generated through fossil fuel or hydroelectric plants. Another problem is that the supply of fissionable uranium-235 is limited. Of all the naturally occurring uranium, only about 0.75 percent is U-235. A vast majority is nonfissionable

U-238. At current usage levels, we'll be out of naturally occurring U-235 in fewer than 100 years. A little bit more time can be gained through the use of breeder reactors (see "Producing plutonium with breeder reactors," later in this chapter). But the amount of nuclear fuel available in the earth is limited, just as the amount of fossil fuels is.

Safety issues: Accidents at Three Mile Island and Chernobyl

Although nuclear power reactors really do have a good safety record, the distrust and fear associated with radiation make most people sensitive to safety issues and accidents. The most serious accident to occur in the United States happened in 1979 at the Three Mile Island plant in Pennsylvania. A combination of operator error and equipment failure caused a loss of reactor core coolant. The loss of coolant led to a partial meltdown and the release of a small amount of radioactive gas. There was no loss of life or injury to plant personnel or the general population.

The outcome was much worse at Chernobyl, Ukraine, in 1986. Human error, along with poor reactor design and engineering, contributed to a tremendous overheating of the reactor core, causing it to rupture. Two explosions and a fire resulted, blowing apart the core and scattering nuclear material into the atmosphere. A small amount of this material made its way to Asia and much of Europe. Hundreds of people died. Many others felt the effect of radiation poisoning. Instances of thyroid cancer, possibly caused by the release of I-131, have risen dramatically in the towns surrounding Chernobyl, and the full effects of this disaster won't be fully known for many years.

Disposal issues: Getting rid of the stuff

The fission process produces large amounts of radioactive isotopes. If you refer to Table 20-2, you notice that some of the half-lives of radioactive isotopes are rather long. Those isotopes are safe after ten half-lives. The length of ten half-lives presents a problem when dealing with the waste products of a fission reactor.

Eventually, all reactors must have their nuclear fuel replenished. And as governments disarm nuclear weapons, they must deal with their radioactive material. Many of these waste products have long half-lives. How can the isotopes be safely stored until their residual radioactivity has dropped to safe limits (ten half-lives)? How can the environment and ourselves and our children for generations to come be protected from this waste? These questions are undoubtedly the most serious problem associated with the peaceful use of nuclear power.

Nuclear waste is divided into low-level and high-level material, based on the amount of radioactivity being emitted. In the United States, low-level wastes are stored at the site of generation or at special storage facilities. The wastes are basically buried and guarded at the sites. High-level wastes pose a much larger problem. They're temporarily being stored at the site of generation, with plans to eventually seal the material in glass and then in drums. The material will then be stored underground in Nevada. At any rate, the waste must be kept safe and undisturbed for at least 10,000 years. Other countries face the same problems. Some nuclear material has been dumped into deep trenches in the sea, but this practice has been discouraged by many nations.

Producing plutonium with breeder reactors

Only the U-235 isotope of uranium is fissionable, because it's the only isotope of uranium that produces the excess of neutrons needed to maintain a chain-reaction. The far more plentiful U-238 isotope doesn't produce those extra neutrons.

The other commonly used fissionable isotope, plutonium-239 (Pu-239), is very rare in nature. But scientists can make Pu-239 from U-238 in a special fission reactor called a *breeder reactor*. Uranium-238 is first bombarded with a neutron to produce U-239, which decays to Pu-239. The process is shown in Figure 20-4.

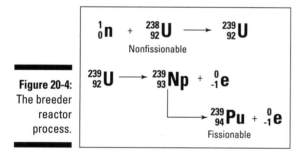

Figure 20-4: The breeder reactor process.

$$^{1}_{0}\text{n} + ^{238}_{92}\text{U} \longrightarrow ^{239}_{92}\text{U}$$

Nonfissionable

$$^{239}_{92}\text{U} \longrightarrow ^{239}_{93}\text{Np} + ^{0}_{-1}\text{e}$$

$$\longrightarrow ^{239}_{94}\text{Pu} + ^{0}_{-1}\text{e}$$

Fissionable

Breeder reactors can extend the supply of fissionable fuels for many, many years, and they're currently being used in France. But the United States is moving *slowly* with the construction of breeder reactors because of several problems associated with them. First, they're extremely expensive to build. Second, they produce large amounts of nuclear wastes. And, finally, the plutonium that's produced is much more hazardous to handle than uranium and can easily be used in an atomic bomb.

Harnessing Nuclear Fusion: The Hope for Tomorrow's Energy

Soon after the fission process was discovered, another process, called *fusion*, was discovered. Fusion is essentially the opposite of fission. In fission, a heavy nucleus is split into smaller nuclei. With fusion, lighter nuclei are fused into a heavier nucleus.

The fusion process is the reaction that powers the sun. On the sun, in a series of nuclear reactions, four isotopes of hydrogen-1 are fused into a helium-4 with the release of a tremendous amount of energy. Here on earth, two other isotopes of hydrogen are used: H-2, called deuterium, and H-3, called tritium. Deuterium is a minor isotope of hydrogen, but it's still relatively abundant. Tritium occurs naturally in minute amounts, but it can easily be produced by bombarding deuterium with a neutron. The fusion reaction is shown in the following equation:

$$\,^3_1H + \,^2_1H \rightarrow \,^4_2He + \,^1_0n$$

The first demonstration of nuclear fusion — the hydrogen bomb — was conducted by the military. A hydrogen bomb is approximately 1,000 times as powerful as an ordinary atomic bomb.

The isotopes of hydrogen needed for the hydrogen bomb fusion reaction were placed around an ordinary fission bomb. The explosion of the fission bomb released the energy needed to provide the *activation energy* (the energy necessary to initiate, or start, the reaction) for the fusion process.

Nuclear fusion as used in a hydrogen bomb may be fine for warfare, but to make the fusion process usable as an energy source for everyday life, the power has to be harnessed, much as it was in nuclear fission. However, developing a fusion power plant has proven to be much more difficult than a fission one.

Overcoming the control issues

The goal of scientists for the last 50 years has been the controlled release of energy from a fusion reaction. If the energy from a fusion reaction can be released slowly, it can be used to produce electricity. It will provide an unlimited supply of energy that has no wastes to deal with or contaminants to harm the atmosphere — simply non-polluting helium. But achieving this goal requires overcoming three problems:

- Temperature
- Time
- Containment

Reaching the right temperature

The fusion process requires an extremely high activation energy. Heat is used to provide the energy, but it takes a *lot* of heat to start the reaction. Scientists estimate that the sample of hydrogen isotopes must be heated to approximately 40,000,000K. (K represents the Kelvin temperature scale. To get the Kelvin temperature, you add 273 to the Celsius temperature. Chapter 3 explains all about Kelvin and his pals Celsius and Fahrenheit.)

But 40,000,000K is hotter than the sun! At this temperature, the electrons have long since left the building; all that's left is a positively charged *plasma,* bare nuclei heated to a tremendously high temperature. Presently, scientists are trying to heat samples to this high temperature through two ways — magnetic fields and lasers. Neither one has yet achieved the necessary temperature.

Maintaining for enough time

Time is the second problem scientists must overcome to achieve the controlled release of energy from fusion reactions. The charged nuclei must be held together close enough and long enough for the fusion reaction to start. Scientists estimate that the plasma needs to be held together at 40,000,000K for about one second.

Containing the reaction

Containment is the major problem facing fusion research. At 40,000,000K, everything is a gas. The best ceramics developed for the space program would vaporize when exposed to this temperature. Because the plasma has a charge, magnetic fields can be used to contain it — like a magnetic bottle. But if the bottle leaks, the reaction won't take place. And scientists have yet to create a magnetic field that won't allow the plasma to leak. Using lasers to zap the hydrogen isotope mixture and provide the necessary energy bypasses the containment problem. But scientists have not figured out how to protect the lasers themselves from the fusion reaction.

Imagining what the future holds

The latest estimates indicate that science is a few years away from showing that fusion can work by reaching the so-called *break-even point,* where the reaction produces more energy than is put in. It will then be another 20 to 30 years before a functioning fusion reactor is developed. But scientists are optimistic that controlled fusion power will be achieved. The rewards are great   an unlimited source of nonpolluting energy.

An interesting by-product of fusion research is the *fusion torch* concept. With this idea, the fusion plasma, which must be cooled in order to produce steam, is used to incinerate garbage and solid wastes. Then the individual atoms and small molecules that are produced are collected and used as raw materials for industry. It seems like an ideal way to close the loop between waste and raw materials. Time will tell if this concept will eventually make it into practice.

Identifying the Effects of Radiation

Radiation can have two basic effects on the body:

- **It can destroy cells with heat.** Radiation generates heat. This heat can destroy tissue, much like a sunburn does. In fact, the term *radiation burn* is commonly used to describe the destruction of skin and tissue due to heat.

- **It can ionize and fragment cells.** Radioactive particles and radiation have a lot of *kinetic energy* (energy of motion — see Chapter 3) associated with them. When these particles strike cells within the body, they can *fragment* (destroy) the cells or *ionize* cell components — turn them into ions (charged atoms) by knocking off an electron. (Flip to Chapter 4 for the full scoop on ions.) Ionization weakens bonds and can lead to the damage, destruction, or mutation of the cells.

Playing hide-and-seek with radon

Radon is a radioactive isotope that's been receiving a lot of publicity recently. Radon-222 is formed naturally as part of the decay of uranium. It's an unreactive noble gas, so it escapes from the ground into the air. Because it's heavier than air, it can accumulate in basements.

Radon itself has a short half-life of 3.8 days, but it decays to Polonium-218, a solid. So if radon is inhaled, solid Po-218 can accumulate in the lungs. Po-218 is an alpha emitter, and, even though this type of radiation is not very penetrating, it has been linked to increased instances of lung cancer. In many parts of the United States, radon testing is performed before selling a house. Commercial test kits can be opened, left in the basement area for a specified amount of time, and then sent to a lab for analysis. The question of whether radon represents a serious problem is still being investigated and debated.

Part V
The Part of Tens

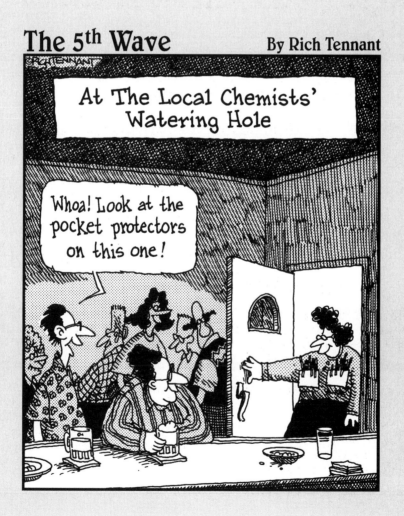

In this part . . .

Chemistry thrives on discovery. And sometimes the discoveries are accidental. The first chapter in this part shows my top-ten favorite accidental discoveries. I also present ten great chemistry nerds and ten terrific tips for passing Chem I. (Study hard and prosper!) I end with a fun chapter of the ten most common chemicals used today to show you the significant role chemistry plays in your everyday life.

Chapter 21

Ten Serendipitous Discoveries in Chemistry

*T*his chapter presents ten stories of good scientists — individuals who discovered something they didn't know they were looking for.

Taking the Measure of Volume

Archimedes was a Greek mathematician who lived in the third century BC. (I know this chapter is supposed to be about scientists and not mathematicians, but back then, Archimedes was as close to a scientist as you could get.) Hero, the king of Syracuse, gave Archimedes the task of determining whether Hero's new gold crown was composed of pure gold, which it was supposed to be, or whether the jeweler had substituted an alloy and pocketed the extra gold. Archimedes knew about density, and he knew the density of pure gold. He figured that if he could measure the crown's density and compare it to that of pure gold, he'd know whether the jeweler had been dishonest. But although he knew how to measure the crown's weight, he couldn't figure out how to measure its volume in order to get the density.

Needing some relaxation, he decided to bathe at the public baths. As he stepped into the full tub and saw the water overflow, he realized that the volume of his body that was submerged was equal to the volume of water that overflowed. He had his answer for measuring the volume of the crown. He got so excited that he ran home naked through the streets, yelling "Eureka, eureka!" ("I've found it!") And this method of determining the volume of an irregular solid is still used today. (By the way, the crown was an alloy, and the dishonest jeweler received swift justice.)

Keeping Rubber Solid

Rubber, in the form of latex, was discovered in the early 16th century in South America, but it gained little acceptance because it became sticky and lost its shape in the heat.

Charles Goodyear was trying to find a way to make the rubber stable when he accidentally spilled a batch of rubber mixed with sulfur on a hot stove. He noticed that the resulting compound didn't lose its shape in the heat. Goodyear went on to patent the *vulcanization process,* which is the chemical process used to treat crude or synthetic rubber or plastics to give them useful properties such as elasticity, strength, and stability.

Right- and Left-Handed Molecules

In 1884, the French wine industry hired Louis Pasteur to study a compound left on wine casks during fermentation — racemic acid. Pasteur knew that racemic acid was identical to tartaric acid, which was known to be *optically active* — that is, it rotated polarized light in one direction or another.

When Pasteur examined the salt of racemic acid under a microscope, he noticed that two types of crystals were present and that they were mirror images of each other. Using a pair of tweezers, Pasteur laboriously separated the two types of crystals and determined that they were both optically active, rotating polarized light the same amount but in different directions. This discovery opened up a new area of chemistry, *optical isomerism,* and showed how important molecular geometry is to the properties of molecules, especially biochemical ones. Pasteur's work was instrumental in the understanding of enzymes and the way they function in the body.

Finding a Shortcut to Color: Artificial Dye

In 1856, William Perkin, a student at the Royal College of Chemistry in London, decided to stay home during the Easter break and work in his lab on the synthesis of quinine. (I guarantee you that working in the lab isn't what my students do during their Easter break!)

During the course of his experiments, Perkin created some black gunk. As he was cleaning the reaction flask with alcohol, he noticed that the gunk dissolved and turned the alcohol purple — mauve, actually. This was the synthesis of the first artificial dye. As luck would have it, mauve was "in" that year, and this dye quickly became in great demand. So Perkin quit school and, with the help of his wealthy parents, built a factory to produce the dye.

Now if this were the entire story, it would've had little effect on history. However, the Germans saw the potential in this chemical industry and invested a great deal of time and resources in it. They began building up and investigating great supplies of chemical compounds, and soon Germany led the world in chemical research and manufacturing.

Dreaming Up the Ring Structure

Friedrich Kekule, a German chemist, was working on the structural formula of benzene, C_6H_6, in the mid-1860s. Late one night he was sitting in his apartment in front of a fire. He began dozing off and, half dreaming, saw groups of atoms dancing in the flames like snakes. Then, suddenly, one of the snakes reached around and made a circle, or a ring. This vision startled Kekule to full consciousness, and he realized that benzene had a ring structure. He stayed up all night working out the consequences of his discovery. Kekule's model for benzene paved the way for the modern study of aromatic compounds.

Discovering Radioactivity

In 1856, Henri Becquerel was studying the *phosphorescence* (glowing) of certain minerals when exposed to light. In his experiments, he'd take a mineral sample, place it on top of a heavily wrapped photographic plate, and expose it to strong sunlight.

He was preparing to conduct one of these experiments when a cloudy spell hit Paris. Becquerel put a mineral sample on top of the plate and put it in a drawer for safekeeping. Days later, he went ahead and developed the photographic plate and, to his surprise, found the brilliant image of the crystal, even though it hadn't been exposed to light. The mineral sample contained uranium. Becquerel had discovered radioactivity.

Finding Really Slick Stuff: Teflon

Roy Plunkett, a DuPont chemist, discovered Teflon in 1938. He was working on the synthesis of new refrigerants. He had a full tank of tetrafluoroethylene gas delivered to his lab, but when he opened the valve, nothing came out. He wondered what had happened, so he cut the tank open. He found a white substance that was very slick and unreactive. The gas had polymerized into the substance now called Teflon. It was used during World War II to make gaskets and valves for the atomic bomb processing plant. After the war, Teflon finally made its way into the kitchen as a nonstick coating for frying pans.

Stick 'Em Up! Sticky Notes

In the mid-1970s, a chemist by the name of Art Frey was working for 3M in its adhesives division. Frey, who sang in a choir, used little scraps of paper to keep his place in his choir book, but they kept falling out. At one point, he remembered an adhesive that had been developed but rejected a couple years earlier because it didn't hold things together well. The next Monday, he smeared some of this "lousy" adhesive on a piece of paper and found that it worked very well as a bookmark — and it peeled right off without leaving a residue. Thus was born those little yellow sticky notes you now find posted everywhere.

Growing Hair

In the late 1970s, Minoxidil, patented by Upjohn, was used to control high blood pressure. In 1980, Dr. Anthony Zappacosta mentioned in a letter published in *The New England Journal of Medicine* that one of his patients using Minoxidil for high blood pressure was starting to grow hair on his nearly bald head.

Dermatologists took note, and one — Dr. Virginia Fiedler-Weiss — crushed up some of the tablets and made a solution that some of her patients applied topically. It worked in enough cases that you now see Minoxidil as an over-the-counter hair-growth medicine.

Speaking of Sweet Somethings

In 1879, a chemist by the name of Constantin Fahlberg was working on a synthesis problem in the lab. He accidentally spilled on his hand one of the new compounds he'd made, and he noticed that it tasted sweet. (Wouldn't the government's Occupational Safety and Health Administration have loved that!) He called this new substance *saccharin*.

In 1965, James Schlatter discovered the sweetness of *aspartame* while working on a compound used in ulcer research. He accidentally got a bit of one of the esters he'd made on his fingers. He noticed its sweetness when he licked his fingers while picking up a piece of paper.

Chapter 22

Ten (Or So) Great Chemistry Nerds

. . . .

In This Chapter

▶ Finding out how some scientists have influenced the field of chemistry

▶ Discovering some great discoveries

▶ Accepting the role of individuals in science

. . . .

Science is a human enterprise. Scientists draw on their knowledge, training, intuition, and hunches. (And as I show you in Chapter 21, serendipity and luck come into play, also.) In this chapter, I introduce you to some scientists whose discoveries advanced the field of chemistry. I had literally hundreds of great scientists to choose from, but following are my top ten . . . plus one.

Amedeo Avogadro

In 1811, the Italian lawyer-turned-scientist Avogadro was investigating the properties of gases when he derived his now-famous law: Equal volumes of any two gases at the same temperature and pressure contain the same number of particles. From this law, the number of particles in a mole of any substance was determined. It was named Avogadro's number. Every chemistry student and chemist knows Avogadro's number. See Chapter 8 if you don't.

Niels Bohr

Niels Bohr, a Danish scientist, used the observation that heated elements emit energy in a set of distinct lines called a *line spectrum* to develop the idea that electrons can exist only in certain distinct, discrete energy levels in the atom. Bohr reasoned that the spectral lines resulted from the transition between these energy levels.

Bohr's model of the atom was the first to incorporate the idea of energy levels, a concept that's now universally accepted. For his work, Bohr received the Nobel Prize in 1922. Check out Chapter 4 for more on Bohr.

Marie Curie

Marie Curie, sometimes called Madame Curie, was born in Poland, but she did most of her work in France. Her husband, Pierre, was a physicist, and both were involved in the initial studies of radioactivity (a term that she coined). Madame Curie discovered that the mineral pitchblende contained two elements more radioactive than uranium. These elements turned out to be polonium and radium. She and her husband shared the Nobel Prize with Henri Becquerel in 1903.

John Dalton

In 1803, John Dalton introduced the first modern atomic theory. He developed the relationship between elements and atoms and established that compounds were combinations of elements. He also introduced the concept of atomic mass.

Unlike many other scientists who had to wait many years to see their ideas accepted, Dalton watched the scientific community readily embrace his theories. His ideas explained several laws that had already been observed and laid the groundwork for the quantitative aspects of chemistry. Not too bad for an individual who started teaching at the age of 12!

Michael Faraday

Michael Faraday made a tremendous contribution to the area of electrochemistry. He coined the terms *electrolyte, anion, cation,* and *electrode.* He established the laws governing electrolysis, discovered that matter has magnetic properties, and discovered several organic compounds, including benzene. He also discovered the magnetic induction effect, laying the groundwork for the electric motor and transformer. Without Faraday's discoveries, I might have had to write this book with a quill pen by lamplight.

Antoine Lavoisier

Antoine Lavoisier is sometimes called the father of chemistry. He was a careful scientist who made detailed observations and planned his experiments. These characteristics allowed him to relate the process of respiration to the process of combustion. He coined the term *oxygen* for the gas that had been isolated by Joseph Priestly in 1774. His studies led him to the *law of conservation of matter,* which states that matter can neither be created nor destroyed. This law was instrumental in helping Dalton develop his atomic theory.

Dmitri Mendeleev

Mendeleev is regarded as the originator of the periodic table, a tool that's indispensable in chemistry. He discovered the similarities in the elements while preparing a textbook in 1869. He found that if he arranged the then-known elements in order of increasing atomic weight, a pattern of repeating properties emerged. He used this concept of *periodic,* or repeating, properties to develop the first periodic table. (Check out Chapter 5 for an illustration of the periodic table used today.)

Mendeleev even recognized that his periodic table had holes where unknown elements should be placed. Based on the periodic properties, Mendeleev predicted the properties of these elements. Later, when gallium and germanium were discovered, scientists found that these elements had properties that were very close to those predicted by Mendeleev.

Linus Pauling

If Lavoisier is the father of chemistry, then Linus Pauling is the father of the chemical bond. His investigations into the exact nature of how bonding occurs between elements were critical in the development of our modern understanding of bonding. His book, *The Nature of the Chemical Bond,* is a classic in the field of chemistry.

Pauling received a Nobel Prize in 1954 for his work in chemistry. He received another Nobel Prize, for peace, in 1963 for his work on limiting the testing of nuclear weapons. He's the only individual to receive two unshared Nobel Prizes. (He's also well known for his advocacy of using megadoses of Vitamin C to cure the common cold.)

Ernest Rutherford

Although Rutherford is perhaps better classified as a physicist, his work on the development of the modern model of the atom allows him to be placed with chemists.

He did some pioneer work in the field of radioactivity, discovering and characterizing alpha and beta particles — and received a Nobel Prize in chemistry for this work. But he's perhaps better known for his scattering experiments, in which he realized that the atom was mostly empty space and that a dense, positive core had to be at the center of the atom, which is now known as the nucleus. Inspired by Rutherford, many of his former students went on to receive their own Nobel Prizes.

Glenn Seaborg

Glenn Seaborg, while working on the Manhattan Project (the atomic bomb project), became involved in the discovery of several of the *transuranium elements* — elements with an atomic number greater than 92. Seaborg came up with the idea that the elements Th, Pa, and U were misplaced on the periodic table and should be the first three members of a new rare earth series under the lanthanides.

After World War II, he published his idea, which was met with strong opposition. He was told that he would ruin his scientific reputation if he continued to express his theory. But, as he said, he had no scientific reputation at that point. He persevered and was proven correct. And he received the Nobel Prize in 1951.

That Third-Grade Girl Experimenting with Vinegar and Baking Soda

This third-grade girl represents all those children who, each and every day, are making great discoveries. They explore the world around them with magnifying glasses. They pry open owl pellets and see what animals the owl ate. They experiment with magnets. They watch while baby animals are being born. They build vinegar-and-baking-soda volcanoes. They discover that science is fun.

They listen when they're told that scientists must keep on trying and that they must not give up. Their parents and teachers encourage them. They aren't told that they can't do science — or if they are, they don't believe it. They ask questions, lots of questions. They love the diversity of science, and they appreciate the beauty of science. These kids may never become professional scientists themselves, but they'll sit at the dining room table someday, laughing and joking with their kids as they help them build vinegar-and-baking-soda volcanoes.

Chapter 23

Ten Terrific Tips for Passing Chem I

In This Chapter

▶ Finding out ways to help your study of chemistry

▶ Exploring techniques that increase exam scores

You may have bought this book to simply discover more about chemistry, but chances are you are or soon will be enrolled in your first year of high school chemistry or first semester of a chemistry course. Naturally, you're interested in maximizing your course grade along with your knowledge. Studying chemistry is different than studying history, and cramming the night before an exam most likely isn't going to result in a good exam grade (so I've been told by students repeating the course). In this chapter, I give you some tips to effectively study chemistry so that you can do well in your chemistry course.

Have a Regular Study Schedule

In all likelihood, you won't get good chemistry grades by cramming right before an exam. Chemistry is one of those subjects that you have to work at in a steady fashion, giving yourself time to think about the material. Try to budget some time every day for your study of chemistry. If you study a bit each day, you won't have to cram the night before the test, and you'll have a much better understanding of the material.

How much time should you devote? The general rule of two hours studying for every hour in class is a pretty good one. Sometimes you may have to spend more; sometimes less. And put away your phone; texting while studying is not an effective way to study.

Strive For Understanding — Don't Just Memorize

When I was in school, many of my classmates tried to simply memorize their way through their chemistry classes. And it didn't work for them, especially when they hit the final exam and they had to try to recall an entire semester's worth of chemistry. I, on the other hand, had such a bad memory that I concentrated on actually learning the material, and I had a much easier and more productive time.

Sure, you will have to memorize some things, but I urge you to strive for understanding. Don't simply memorize the fact that the size of atoms decreases when going from left to right in a period; learn *why*. That way, when faced with a similar situation, such as the trends in ionization potentials, you can reason your way through. The less you have to memorize, the less you have to forget in the middle of an exam.

Also, know what will be supplied on the test and what won't. What constants will you be expected to know? Do you need to memorize the entire periodic table? (Okay, that was a joke!)

Practice by Doing the Homework

Chemistry is not a spectator sport. I've had many, many students over my 40 years of teaching tell me that those problems looked so easy when I worked them, but they couldn't do them on the exam. They weren't successful because they never practiced. Ever watch professional athletes or musicians perform? They make it look easy. But behind those performances is practice, practice, and more practice.

The same is true in chemistry; to be good, you must practice. Homework is practice. The more homework problems you do, the better your performance on an exam will be. If you have trouble with your homework and can't finish it, ask your teacher or a tutor to meet with you before the due date. And not only should you do all assigned homework (obviously important if you want a good grade!), but you also should find additional problem sets and do them. Practice, practice, practice.

Get Help from Additional Resources

When I'm preparing a lesson on an unfamiliar topic, I may refer to different resources to help me reach a level of understanding that will allow me to explain the concept to my students. I use several different books to see the approach that different authors take. I use the Web to find current information or additional problem sets. If I have a colleague who is familiar with the topic, I talk to him or her.

Preparing a one-hour lecture takes several hours. You need to put forth the same amount of time when studying chemistry. No matter how good the author of your textbook is, some things in the text won't be crystal clear. That's where you can refer to other resources, like this book, to help you figure out what the heck your textbook's author is trying to say. You likely don't have an extensive chemistry library (probably this guide and your textbook), so go to the library and work with some other text. Do an online search for topics you are having trouble with. Look at several sources from several angles. Sure, doing so takes time, but a lot less time than repeating the course.

Read the Material before Class

Chemistry teachers often lecture over new material without expecting you to study the topic beforehand. However, skimming the text before going to class helps set the stage for the lecture. I tell my students to spend about 15 minutes reading ahead in their text before coming to class. Just read for general understanding, don't work problems (that comes later). Simply try to get an overview of the topic that your teacher will be covering. This prior understanding really will make the lecture much more meaningful. After class, you can reread in depth; before class, you are looking for that big picture.

If, on the other hand, your teacher expects you to read the material in depth before coming to class, then you had better read it and even attempt the problems.

Take Good Notes

I've been tempted to buy a T-shirt that says, "If my lips are moving, you should be writing!" Theoretically, you should be able to just read the textbook, work the exercises, and ace the exams. However, it never seems to work that way. Your instructor (hopefully) won't simply be reading from the text. She will be explaining things in a different way than the textbook author does, which is very valuable.

Your teacher will also be explaining why things happen as they do. Again, this information is valuable because you're going beyond simple memorization. Taking good notes allows you to capture those explanations and techniques. Later when you're studying for a quiz or test, the notes can help jog your memory; the better and more complete the notes, the more useful they become, which means the better you can do on that exam.

Recopy Your Lecture Notes

Using a laptop to take notes in a chemistry class is difficult because of all the special symbols, structures, formulas, and so on. Therefore, most students take handwritten notes, and those notes tend to be sloppy because they're written fast. I've also been told that some students actually fall asleep in class (but not in my class), and their notes trail off.

Sloppy notes are difficult to study from, so make sure you recopy your notes to make them more useful. You can take a little more time drawing those structures and writing those formulas. Try to recopy your notes as soon as possible after the lecture, within 24 hours if possible, as a reinforcing technique to help you comprehend difficult topics. And if something isn't clear, you can use your text and/or other sources to supplement and resolve any misconceptions.

You can even take the time to type the notes on your computer so that you can quickly search through them for certain topics later on when you're studying. Drawing may still be an issue in typed notes, but you can use drawing software or just keep hand-drawn figures separately.

Ask Questions

The only stupid question is the one you didn't ask. If you don't understand something in class, ask questions as soon as possible. Normally, the subject won't get any clearer if you wait, and you may not be able to understand important information that builds on it. If you're shy about asking a question in class, go to your instructor's office during office hours. Have him explain it again. Maybe a different approach will help you. Most teachers and professors entered the field because they enjoy helping people, so please give your teacher an opportunity to help. You are (or someone else is) paying a great deal of money for your instructor's time, so make sure you get your money's worth!

Get a Good Night's Sleep before Exams

If you've kept up with your studying and homework, then all that should be left to do the night before an exam is a little brief review. Learning a lot of chemistry the night before the exam is difficult. Rather, the night before, you want to glance over your notes a few times and spend at most an hour studying.

The best thing you can do is to get a good night's sleep so that you're refreshed and relaxed for the exam. You need to be able to think your way through the chemistry test, which is difficult if you keep falling asleep. All-nighters are a no-no.

Pay Particular Attention to Details

If you know the material, you can still get a bad grade by making careless errors on your exams. Checking to make sure you have shown the correct charges on charged species (ions, subatomic particles, and so on) and that you have reported your answer to the correct number of significant figures really helps in reducing the number of lost points. You also want to make sure you haven't skipped any problems, you've filled your answer sheet out properly, and so on. Silly careless mistakes can cost you easy points.

Chapter 24

The Top Ten Industrial Chemicals

*I*n the final analysis, chemistry is about chemicals. In school you may carry out a reaction with a few grams of a chemical; in industry tons of the same chemical may be used in the same reaction. And in industry a lot of money is made from actually very few chemicals. I know that I wondered about industrial chemistry when I was starting out in chemistry classes, including what went on in that field and what chemicals were used on a large scale, so in this chapter I tell you about the ten most commonly produced chemicals. I don't list the amounts produced, because they change from year to year, but all the chemicals listed are produced in excess of 100 million metric tons. Here you can see how much chemistry impacts your daily life.

Sulfuric Acid (H_2SO_4)

No matter what the year, sulfuric acid heads the list as the #1 produced chemical worldwide. The major use of sulfuric acid is in the production of fertilizers — ammonium sulfate and superphosphate. However, sulfuric acid is also used in other products, including the following of its function:

✔ Detergents

✔ Lead-acid automobile batteries

✔ Other chemicals such as hydrochloric acid, dye, explosives, pigments, and drugs

Sulfuric acid is also used as a reactant during the manufacture or processing of certain goods. Following are some examples of its function:

✔ To remove impurities during petroleum refining

✔ To remove metallic oxides before electroplating and galvanizing metal

✔ To remove water during certain chemical reactions

✔ To act as a reactant in the manufacture of rayon and nitroglycerine

Nitrogen (N₂)

Nitrogen is largely an *inert* gas because it's commonly used as a *blanketing gas,* which means it protects oxygen-sensitive materials from contact with the air. Following are some of the many industrial uses of liquid nitrogen:

✔ To quickly freeze substances for processing. For example, it's commonly used to freeze old tires in order to make them easier to shred for recycling purposed.

✔ To manufacture steel and other metals.

✔ To cool concrete, improving the properties of the building material.

✔ To freeze soggy ground, making construction easier.

✔ To cool chemical reactors, allowing chemical engineers to more effectively control side reactions.

The following industries also use nitrogen and liquid nitrogen:

✔ The food industry utilizes the quick cooling aspect to minimize cell damage from ice crystals that commonly form during the normal freezing process. Another use in food service: It's used in refrigerated trucks to minimize the contact of the food with air.

✔ The healthcare industry uses it to freeze blood and tissue samples as well as in cryo-surgery to destroy tissue, such as warts.

And my college chemistry club uses it to make ice cream. No churning here; it's ready in less than a minute!

Ethylene (C₂H₄)

Ethylene is one of the major feed stocks for the chemical industry, especially the plastics industry. You may be surprised to see how versatile this chemical is. It's used in these ways:

✔ To produce ethylene glycol (antifreeze), styrene (used to make polystyrene for use as packing and insulation), and polyethylene, one of the most widely used plastics. In fact, about half of the ethylene produced is used to make the various types of polyethylene.

✔ To make ethanol for industrial uses (by law, ethanol for human consumption must be produced by fermentation).

✔ To produce polyester (like for the polyester leisure suits of the 1970s).

✔ To produce synthetic rubber.

Oxygen (O_2)

One of oxygen's main roles is in the following combustion processes:

✔ Commercially produced oxygen is used in oxyacetylene and oxyhydrogen welding torches.

✔ Oxygen is used in the steelmaking industry to help burn off impurities in the molten ore. About a ton of oxygen is required per ton of steel produced!

✔ Liquid oxygen (LOX) is used as an oxidizing agent in missiles and rockets. The oxygen tank that is used to launch the space shuttle holds about 550,000 liters of liquid oxygen.

Oxygen is also used in chemical industries to break down hydrocarbons (compounds of just carbon and hydrogen) into smaller hydrocarbon products such as ethylene, propylene, and acetylene, which are in turn used to produced plastics, paints, and other products.

Propylene (C_3H_6)

Propylene's major use is as an intermediate in the production of other chemical compounds, like the following:

✔ In the production of polyethylene needed to produce synthetic fibers for indoor/outdoor carpets

✔ In the production of propylene glycols for auto brake fluid, detergents, and paints

✔ In the production of polyurethane for rigid foam insulation

✔ In the production of various types of ABS plastics used in telephones and auto trim parts

Chlorine (Cl_2)

Chlorine has many uses, including the following:

- ✔ To produce consumer and industrial products such as plastics, pharmaceuticals, dyes, household cleaners (including bleach and other disinfecting agents), insecticides, and textiles
- ✔ To treat water in multiple ways
  - To largely eliminate waterborne microbiological infections during water purification in water treatment plants in the United States
  - To kill bacteria in swimming pools (sodium hypochlorite produced from chlorine is used)
- ✔ To act as a major reactant in the production of bulletproof vests, computer chips, and auto parts

Ethylene Dichloride ($C_2H_2Cl_2$)

The major use of ethylene dichloride is in the production of polyvinyl chloride (PVC). Those white plastic pipes used to carry water underground and throughout your home are PVC, which is a mainstay of the construction industry.

Ethylene dichloride is also used in the production of polystyrene, another useful synthetic polymer. Ethylene dichloride is used in the production of

- ✔ Certain dry cleaning fluids
- ✔ Flooring
- ✔ Shower curtains
- ✔ Synthetic rubber

Phosphoric Acid (H_3PO_4)

About three-quarters of the phosphoric acid produced worldwide is used in the production of synthetic phosphate fertilizers. Other uses for phosphoric acid include

- ✔ **Food additive:** It's added in the food industry as a food pH adjuster (in colas, for example), as a clarifying agent, and as a preservative.

✔ **Rust remover:** It acts as a rust converter, converting the ferric oxide (Fe_2O_3) to ferric phosphate ($FePO_4$), which can then be easily scrubbed off. Phosphoric acid for this purpose is commonly sold as a gel called naval jelly.

Ammonia (NH_3)

Well over half of the ammonia produced worldwide is used in agriculture.

✔ It's used to produce liquid fertilizers that contain ammonia, ammonium nitrate, and urea. It is also used in the production of ammonium nitrate fertilizer.

✔ It's used in the production of cotton defoliants, stripping leaves to make the cotton easier to pick.

✔ It's used to make antifungal agents for certain fruits.

It is used in the production of other chemicals and products, including

✔ Nitric acid

✔ Certain dyes

✔ Sulfa drugs

✔ Cosmetics

✔ Vitamins

✔ Certain synthetic textiles, such as rayon and nylon

✔ Household cleaners, such as glass cleaners

Additionally, ammonia is used by several industries:

✔ As a complexing agent in the mining and metal manufacturing industries

✔ As a refrigerant in industrial refrigeration

✔ As a curing and protective agent in the leather industry

Sodium Hydroxide ($NaOH$)

In industry when a strong base is required, sodium hydroxide is the one. It's put to many uses in a variety of industries. Following are some of its uses:

✔ Used by the petroleum industry to increase the pH of drilling mud, making it more viscous

- ✔ Used by some countries to help remove sulfur impurities in low-grade crude oil

- ✔ Used in digestion and bleaching of wood fibers by the paper-making industry

- ✔ Used for the decomposition of road kill and, if you can believe *CSI*, to get rid of human remains

- ✔ Used in soap making

- ✔ Used in the production of biodiesel

- ✔ Used as an industrial cleaning agent, especially in the degreasing of equipment, and is used in the home as oven and drain cleaners

- ✔ Used by the food industry in the making of hominy, Chinese noodles, and German pretzels

Glossary

• •

absolute zero: The temperature at which all molecular motion stops, 0 Kelvin.

acid: A compound that is a proton (H^+) donor.

acidic: A solution whose pH is less than 7.

acid rain (deposition): Rain that has a pH in the acid range due to pollutants.

activation energy: The minimum amount of energy that must be supplied in order to start a chemical reaction.

activity series: A list of the metals in order of decreasing ease of oxidation.

actual yield: The amount of product actually formed in a chemical reaction.

alkali metals: The elements in the IA (1) family on the periodic table.

alkaline earth metals: The elements in the IIA (2) family on the periodic table.

alpha particle: Essentially a helium nucleus (two protons and two neutrons).

amorphous solid: A solid that lacks extensive ordering of the particles.

amphoteric: A substance that acts either as an acid or a base depending on what it's combined with.

amplitude: The height of a wave.

amu: An atomic mass unit, $\frac{1}{12}$ of the mass of a C-12 nucleus.

angular momentum quantum number (*l*): This number describes the shape of an orbital.

anions: Ions that have a negative charge.

aqueous solution: A mixture in which the solvent is water.

atom: The smallest particle of matter that represents a particular element.

atomic number (*Z*): The number of protons in the nucleus.

atomic orbital: The volume of space in which you're most likely to find a specific electron in an atom.

aufbau principle: States that the electrons in an atom fill the lowest energy levels first.

Avogadro's law: States that the volume of a gas and the number of moles of gas are directly related if the temperature and pressure are held constant.

Avogadro's number: The number of particles (atoms, ion, molecules) in a mole; it is numerically equal to 6.022×10^{23}.

barometer: An instrument that measures atmospheric pressure.

base: A compound that is a proton (H^+) acceptor.

basic: A solution with a pH greater than 7.

beta particle: Essentially an electron.

binary compound: A compound composed of only two elements.

biological oxygen demand (BOD): The amount of dissolved oxygen needed to oxidize the biological material in water.

boiling: The process of going from a liquid state to a gaseous state.

boiling point (bp): The temperature at which a liquid boils; also the temperature at which the vapor pressure of the liquid equals atmospheric pressure.

bond order: Relates the bonding and antibonding electrons in a molecular orbital and is equal to (# electrons in bonding MO's – # electrons in antibonding MO's)/2.

Boyle's law: States that an inverse relationship exists between the volume and pressure of a gas if the temperature and amount are held constant.

buffers: Solutions that resist a change in pH when either an acid or a base is added to them.

calorie: The amount of energy needed to raise the temperature of 1 gram of water 1 degree Celsius.

calorimetry: A laboratory technique used to measure the amount of heat released or absorbed during a chemical or physical process.

capillary action: The spontaneous rising of a liquid in a narrow tube against the force of gravity.

catalyst: A substance that speeds up a reaction and is (at least theoretically) recoverable at the end of the reaction in an unchanged form.

cations: Ions that have a positive charge.

Charles's law: States that a direct relationship exists between the volume and Kelvin temperature if the pressure and amount of the gas are held constant.

chemical equilibrium: Established when two exactly opposite reactions are taking place at the same time in the same place and with equal rates of reaction.

colligative properties: Properties of solutions that depend only on the number of solute particles present and not on the type of solute.

colloids: Homogeneous mixtures in which the solute diameters are between those of solutions and suspensions.

combined gas equation: Relates the temperature, pressure, and volume of a gas, assuming the amount is held constant.

combustion reaction: A reaction in which a chemical species combines rapidly with oxygen and usually emits heat and/or light.

compounds: Pure substances composed of two or more different elements.

concentrated: A qualitative term that describes a solution with a relatively large amount of solute in comparison to the amount of solvent.

concentration: The measure of the amount of solute dissolved in a solution.

conjugate acid-base pair: A pair of compounds (one an acid and one a base) that differ by only a single H^+.

continuous spectrum: A spectrum of light in which all wavelengths of light are present.

coordinate covalent bond: A covalent bond between two atoms in which one atom has furnished both electrons for the bond.

covalent bond: A bond in which one or more electron pairs are shared between two atoms.

critical point: The point on the phase diagram beyond which the gas and liquid phases of a substance are indistinguishable from each other.

crystal lattice: A three-dimensional structure that crystalline solids occupy.

crystalline solids: A solid in which the particles are arranged in a very regular ordering called a crystalline lattice.

Dalton's law: States that in a mixture of gases, the total pressure is the sum of the pressures of the individual gases.

decomposition reaction: A reaction in which a compound breaks down into two or more simpler substances.

dilute: A qualitative term describing a solution that has a relatively small amount of solute compared to the amount of solvent.

dipole: A molecule in which one end is negative and the other end is positive.

dipole-dipole interaction: An intermolecular force that occurs between polar molecules.

double-displacement reaction (metathesis): A reaction in which at least one insoluble product is formed when mixing two solutions.

effective nuclear charge: The net attraction to the nucleus that an electron experiences taking into effect the shielding effect that the other electrons contribute.

electrolyte: A substance that conducts an electrical current when melted or dissolved in water.

electromagnetic spectrum: The range of radiant energy composed of gamma rays, X-rays, and so on.

electron: The subatomic particle that has a negative charge and very little mass.

electron affinity: The energy change that results from adding an electron to a gaseous atom or ion.

electron capture: A radioactive decay mode in which an electron from the 1s orbital is captured by the nucleus.

electron cloud: A volume of space in which the probability of finding an electron is high. (Also called *electron density.*)

electron configuration: A condensed method of representing the pattern of electrons in an atom.

electronegativity: A measure of the attractive force that an atom has on a pair of bonding electrons.

empirical formula: A chemical formula of a compound that indicates what atoms are present and the simplest whole-number ratio of the elements.

endothermic: A type of reaction that absorbs energy from its surroundings.

endpoint: The point of a titration at which an indicator signals that an equivalent amount of titrant as substance being titrated has been added.

enthalpy change (ΔH): The heat gained or lost by a system during constant pressure conditions.

exothermic: A type of reaction that gives off heat to its surroundings.

families: The vertical columns on the periodic table. (Also called *groups.*)

frequency *(v)*: The number of waves that pass by a reference point per second.

gamma emission: A radioactive decay mode in which high-energy, short-wavelength photons are emitted from the nucleus.

gas: A state of matter that has no definite shape or volume.

Gay-Lussac's law: Describes the direct relationship between the pressure of a gas and its Kelvin temperature if the volume and amount of the gas are held constant.

Graham's law: Shows that the speed of gas diffusion or effusion is inversely proportional to the square root of the gas's molar masses.

greenhouse effect: The warming of the atmosphere due to the absorption of radiant energy by certain gases.

groups: The vertical columns on the periodic table. (Also called *families.*)

half-life: The amount of time that it takes for a substance to decay to exactly one-half of its initial concentration.

halogens: The elements in the VIIA (17) family on the periodic table.

heat capacity: The amount of heat needed to change the temperature of a substance 1K.

heat of vaporization: The amount of heat needed to change a liquid into a gas.

Henry's law: States that the solubility of a gas increases with the increasing partial pressure of the gas.

Hess's law: States that if a reaction occurs in a series of steps, then the enthalpy change for the overall reaction is simply the sum of the enthalpy changes of the individual steps.

Hund's rule: States that electrons add to orbitals of the same energy, half-filling them all before the electrons spin pair.

hybrid orbitals: Atomic orbitals formed as a result of the mixing of the atomic orbitals of the atoms involved in a covalent bond.

hydrogen bonding: A strong dipole-dipole intermolecular force that results from a hydrogen atom bonded to an oxygen, nitrogen, or fluorine on one molecule being attracted to an oxygen, nitrogen, or fluorine atom on another molecule.

ideal gas: A gas that obeys the five postulates of the kinetic molecular theory of gases.

ideal gas equation: Relates the Kelvin temperature, the pressure, the volume, and the amount of a gas; it has the mathematical form of $PV=nRT$.

indicators: Compounds added to the substance being titrated that change color to signal the endpoint.

inner transition elements: The two horizontal groups that are pulled out of the body of the periodic table.

intermolecular forces: Attractive or repulsive forces between molecules.

ion-dipole interaction: Intermolecular forces between an ion and a polar molecule.

ionic bond: A bond resulting from a metal reacting with a nonmetal; the metal loses electrons, forming a cation, while the nonmetal gains electrons, forming anions. The attractive force between the unlike ions is the ionic bond.

ionic equation: Shows the soluble reactants and products in the form of ions.

ionic solids: Solids with crystal lattices composed of ions held together by the attraction of the charges of the ions.

ionization energy: The energy needed to completely remove an electron from a gaseous atom.

isoelectronic: Having the same electronic configuration.

isotopes: Atoms of the same element that have different numbers of neutrons.

joule: The SI unit for energy.

kinetic energy: Energy of motion.

kinetic molecular theory: A model that attempts to describe the properties of gases at the microscopic level.

law of conservation of matter: Says that matter is neither created nor destroyed in ordinary chemical reactions.

Lewis structure: A structural formula that represents the elements and their valence electrons.

limiting reactant: The reactant that is first totally consumed in a chemical reaction.

line spectrum: A series of fine lines representing the wavelengths of photons characteristic of a particular element.

liquid: A state of matter that has a definite volume but no definite shape.

London smog: A gaseous atmospheric mixture of fog, soot, ash, sulfuric acid, and sulfur dioxide.

LUST: A pollution term for leaking underground storage tanks used for the storage of gasoline or other liquids.

magnetic quantum number (ml): Describes the orientation of the orbital around the nucleus.

main-group elements: The groups on the periodic table that are labeled with an A.

manometer: An instrument used to measure the pressure of a confined gas.

mass number: For an element, the sum of its protons and neutrons.

mass (weight) percent: The mass of the solute divided by the mass of the solution and then multiplied by 100%.

mass (weight)-volume percent: The mass of the solute divided by the volume of the solution and then multiplied by 100%.

melting point (mp): The temperature at which the solid state of a substance converts to the liquid state at atmospheric pressure.

metallic bonding: A type of bonding in metals in which the electrons of each atom are delocalized and free to move throughout the entire solid.

metalloid: A group of elements that have properties of both metals and nonmetals.

metals: Elements that are malleable, ductile, and good conductors. They tend to lose electrons in chemical reactions. Mercury is the only liquid metal; the rest are solids.

metathesis reaction: A reaction in which at least one insoluble product is formed from the mixing of two solutions.

molality (m): A solution concentration unit that is defined as the moles of solute per kilogram of solvent.

molar heat capacity (S): The amount of heat needed to change the temperature of 1 mole of a substance 1K.

molarity (M): A solution concentration unit that is defined as the number of moles of the solute per liters of solution.

mole (mol): The number of particles in exactly 12 g of C-12. At the microscopic level there are 6.022×1023 particles/mol, and at the macroscopic level a mole is the number of grams in the molar mass of a substance.

molecular equation: An equation that shows all reactants and products in the undissociated form.

molecular formula: Shows what elements are in the compound and the actual number of each. Also referred to as the true or actual formula.

molecular orbital (MO) theory: Describes covalent bonding as the combination of atomic orbitals to form molecular orbitals that encompass the entire molecule.

molecule: A covalently bonded compound.

net-ionic equation: An equation in which the spectator ions are not shown. Only the chemical species involved in the chemical reaction are shown.

network solid: Covalently bonded substances whose crystal lattice is extremely large.

neutral: A solution with a pH of 7.

neutralization reactions: Acid-base reactions in which an acid reacts with a base to give a salt and water.

noble gases: The VIIIA (18) group on the periodic table. They tend to be unreactive due to their filled valence shells.

nonelectrolytes: Substances that don't conduct electricity when melted or dissolved in water.

nonmetals: The elements that generally have properties the opposite of metals. They tend to gain electrons during chemical reactions.

nonpoint sources: Pollution sources that are diffuse in nature.

nonpolar covalent bond: A bond in which the bonding electrons are shared equally between two atoms.

nucleus: The dense central core holding the protons and neutrons.

octet rule: States that during chemical reactions, atoms lose, gain, or share electrons in order to achieve a filled valence shell, to complete their octet of eight electrons.

orbital (wave function): A quantum mechanical description of the location of an electron in an atom.

osmosis: The passing of solvent molecules through a semipermeable membrane.

osmotic pressure: The amount of pressure that must be exerted on a solution in order to stop osmosis.

oxidation: The loss of electrons.

oxidation numbers: Bookkeeping numbers that allow chemists to do things like balance redox equations.

oxidizing agent: The reactant being reduced.

particulates: Small, solid particles suspended in the air.

pascal: The SI unit of pressure.

periods: Horizontal groupings on the periodic table.

phase changes: Changes of state.

phase diagram: A graphical representation of the relationship of the states of matter of a substance to temperature and pressure.

photochemical smog: An atmospheric gas produced when sunlight initiates certain chemical reactions involving unburned hydrocarbons and oxides of nitrogen.

pi (π) bonds: Bonds resulting from the overlap of atomic orbitals above and below a line connecting the two nuclei.

point sources: Pollution sources that have a definite identifiable source.

polar covalent bond: A bond in which the bonding electron pairs are unequally shared.

positron: Essentially an electron that has a positive charge.

potential energy: Stored energy.

precipitate: An insoluble material that forms in a solution from ions present.

precipitation reactions: Reactions involving the formation of an insoluble precipitate from the mixing of two soluble compounds.

pressure: Force per unit of surface area.

primary sewage treatment: The first stage of the purification process that removes sewage from water, involving settling and filtration to remove large particles.

principal quantum number *(n):* Describes the size of the orbital and relative distance from the nucleus.

proof: Twice the volume percent of an aqueous ethyl alcohol solution.

quantized: A term for an atom with only certain distinct energies associated with its state.

quantum numbers: Describe each electron in an atom; they tell the orbital size, shape, orientation in space and the spin of each electron.

radioactivity: The spontaneous decay of an unstable isotope to a more stable one.

reactants: The starting material in a chemical reaction.

reaction intermediate: A substance that is formed but then consumed during the reaction.

reaction mechanism: The sequence of individual reactions that occur in an overall reaction.

redox reactions: Chemical reactions in which electrons are gained and lost.

reducing agent: The reactant undergoing oxidation in a redox reaction.

reduction: The gain of electrons.

resonance: A way of describing a molecular structure that cannot be represented by a single Lewis structure. Several different Lewis structures are used, each differing by the position of electron pairs.

reverse osmosis: A process that takes place when the pressure on the solution side of a semipermeable membrane exceeds the osmotic pressure and solvent molecules are forced back through the membrane.

saturated solution: A solution in which the maximum amount of solute per given amount of solvent at a certain temperature has been dissolved.

secondary sewage treatment: The second stage of the water purification process that removes sewage from water, using bacteria and other microorganisms to decompose the organic material in wastewater.

semipermeable membrane: A thin porous film that allows the passage of solvent molecules but not solute particles.

shells: The various energy levels at different distances from the nucleus in which electrons in an atom are located.

SI system: The system of units used in science; it is related to the metric system.

sigma bonds: Bonds with the orbital overlap on a line connecting the two nuclei.

single-displacement (replacement) reactions: Reactions in which atoms of an element replace the atoms of another element in a compound.

solid: A state of matter than has both a definite shape and volume.

solute: The component of a solution that is present in the smallest amount.

solution: A homogeneous mixture composed of a solvent and one or more solutes.

solvation: The formation of a layer of solvent molecules surrounding a solute particle.

solvent: The component of a solution that is present in the largest amount.

specific heat capacity *(s):* The quantity of heat needed to raise the temperature of 1 g of a substance 1K.

spectator ion: Anion directly involved in a chemical reaction that maintains electrical neutrality of the solution.

speed of light: The speed that light travels in a vacuum, 3.00×10^8 m/s.

spin quantum number: Indicates the direction the electron is spinning.

standard enthalpy of formation (ΔH_f°): The change in enthalpy when 1 mole of the compound is formed from its elements when all substances are in their standard states.

stoichiometry: The calculation of amount (mass, moles, particles) of a substance in a chemical reaction.

strong acid: A proton donor that completely ionizes in water.

strong base: A proton acceptor that contains the hydroxide ion and completely ionizes in water.

strong electrolytes: Electrolytes that completely dissociate or ionize in water.

structural isomers: Compounds that have the same molecular formula but differ in how the groups are attached to each other.

sublimation: A change of state in which a substance goes directly from the solid state to the gaseous state without becoming a liquid.

subshells: Within the electron shells, orbitals of slightly different energies in which electrons are grouped.

supersaturated solution: A solution in which more than the maximum amount of solute has been dissolved in the solvent.

surface tension: The amount of force required to break through the molecular layer at the surface of a liquid.

surroundings: In thermodynamics, a term that represents the rest of the universe that is being affected by some type of change.

suspension: A heterogeneous mixture whose particles are large (>1,000 nm).

system: The thermodynamics term for the part of the universe that you are studying.

tertiary sewage treatment: The third stage of the process of water purification, using chemicals to remove fine particles, nitrates, and phosphates from the wastewater.

theoretical yield: The maximum amount of product that can be formed during a chemical reaction.

thermochemistry: The part of thermodynamics that deals with changes in heat that take place during chemical reactions.

thermodynamics: The study of heat and its changes.

titrant: The solution in a titration that has a known concentration.

titration: A laboratory technique in which a solution of known concentration is used to determine the concentration of an unknown solution.

transition elements: The elements categorized as B groups on the periodic table.

transmutation: A nuclear reaction in which an element is created by another one.

triple point: The combination of temperature and pressure on a phase diagram in which all three states of matter of a substance can exist.

Tyndall effect: Exhibited when a light is shown through a colloid and the light beam is visible due to the reflection of light from the large colloid particles.

unit cells: The repeating units in a crystal lattice.

unsaturated solution: A solution with less than the maximum amount of solute dissolved in a given amount of solvent at a given temperature.

valence bond theory: Describes covalent bonding as the overlap of atomic orbitals to form a new type of orbital, a hybrid orbital.

valence electrons: The electrons (normally only in s and p orbitals) that are in the outermost energy level.

van der Waals equation: A modification of the ideal gas equation to compensate for the behavior of real gases.

van 't Hoff factor: The ratio of moles of solute particles formed to moles of solute particles dissolved in solution.

vapor pressure: The pressure exerted by the gaseous molecules that are in contact with a liquid in a closed container.

viscosity: The resistance to flow of liquids.

volume percent: For a solution, the volume of the solute divided by the volume of the solution, with the result multiplied by 100%.

VSEPR theory: Predicts molecular geometry by considering that the valence electron pairs around a central atom try to maximize the distance from each other in order to minimize repulsive forces.

wave function: A mathematical description of the electron's motion.

wavelength (λ): The distance between two identical points on a wave.

weak acid: A proton donor that only partially ionizes in water.

weak base: A proton acceptor that only partially ionizes in water.

weak electrolytes: An electrolyte that only partially ionizes in water.

Index